T0325580

ROBUST DESIGN METHODOLOGY FOR RELIABILITY

ROBUST DESIGN METHODOLOGY FOR RELIABILITY

EXPLORING THE EFFECTS OF VARIATION AND UNCERTAINTY

Edited by

Bo Bergman, Jacques de Maré, Sara Lorén and Thomas Svensson

Chalmers University of Technology,
Sweden

A John Wiley and Sons, Ltd., Publication

Library of Congress Cataloging-in-Publication Data

Record on File

A catalogue record for this book is available from the British Library.

ISBN 978-0-470-71394-5 (H/B)

Typeset in 11/13pt Times by Aptara Inc., New Delhi, India
Printed and bound in Great Britain by CPI Antony Rowe, Chippenham, Wiltshire.

Contents

PART Three   MODELLING   113

Preface

This book is the result of a collaboration between researchers and practitioners with an interest in reliability but with different backgrounds, for example quality sciences, mathematical statistics, and fatigue, and industrial experience of both general reliability problems and specific fatigue type problems.

We have observed that many current reliability techniques suggested in typical text books on reliability are not quite in line with the practical problems encountered when designing for reliability as they are usually based on field failure occurrences, i.e. on information obtainable only after the designs have been realized. We have also found that most reliability problems are related to variation and uncertainty. Thus, to design for reliability this should be addressed directly, rather than indirectly utilizing failure rates based on field data, which are often of questionable quality. Moreover, no really good design rules support reliability improvement activities.

With this selection of essays addressing reliability problems in a more constructive manner we take some steps towards a more comprehensive theory of design for reliability. More specifically, we utilize Robust Design Methodology, suggested by Taguchi and promoted in many 'Design for Six Sigma' frameworks. In essence, Robust Design Methodology is a systematic search for design solutions insensitive to variation in usage, manufacturing and deterioration of products as well as to varying system environments. Specifically, we utilize these ideas for design for reliability purposes. We also include uncertainties for which the system solution should be insensitive. These uncertainties include uncertainties on system and model assumptions on which design solutions are based. Furthermore, in contrast to the prevailing theories of robustness we advocate robustness studies already in the early phases of product development in order to support creativity and innovativeness in the design for reliability.

Contrary to most reliability texts, we promote the study of variation and uncertainty as a basis for reliability work. In Davis (2006)[1] a similar framework is described and Clausing also advocated a related approach long ago, see Clausing (2005)[2]. Robust Design Methodology (RDM) is a core technique in our approach. In most texts, the RDM techniques are initiated from Failure Mode and Effects Analysis (FMEA), which was suggested in the sixties and which nowadays has widespread application in industry. As a complement, or possibly, as a

---

[1] Davis, T. P. Science, engineering, and statistics. *Applied Stochastic Models in Business and Industry*, **22**: 401–430, 2006.

[2] Clausing, D. Operating window – an engineering measure for robustness. *Technometrics*, **46**(1): 25–29, 2004.

replacement of FMEA, we suggest an enhancement of that technique – what we call VMEA – Variation Mode and Effects Analysis. In VMEA, sources of variation and uncertainties affecting important outputs are identified and assessed. In our approach, we go immediately to the root causes of failures, i.e. to sources of variation and uncertainties. When important areas for improvement are identified we suggest the utilization of Robust Design Methodology.

Today, many companies use Six Sigma as their strategic improvement initiative. However, to obtain sustainable effects of initiatives of this kind, the design process also has to be addressed. Such initiatives often go under the label 'Design for Six Sigma' (DFSS). However, in many DFSS descriptions, the reliability perspective is not readily described. With the techniques we advocate these relations are made more explicit and clear illustrations from fatigue problems will be given.

For the future development of the field, textbooks elaborating and simplifying the concepts suggested in this directed collection of essays are required. It is our intention to produce such textbooks and we hope to be able to start a new era of reliability improvement work. The evolution of reliability theory may be described in different ways – one way is to separate a number of eras, for example:

1. The reactive era, when reliability improvements were initiated due to problems that had arisen. Many very attractive reliability improvement techniques, which however not always succeeded, were developed; e.g. in mechanical reliability ideas of safe life, fail safe and damage tolerance.
2. The quantitative era, when description of failure processes using statistical concepts, with books by Barlow and Proschan[3] as highlights. An interesting shift in emphasis is the transition from a focus on component reliabilities to a more system-oriented reliability approach.
3. The condition monitoring era, initiated by Nowlan and Heap (1978)[4] and still being developed in many industries.

Now it is time for a new era in reliability thinking, where the design efforts are addressed and, specifically, the generation of failure proneness in the design phase is addressed through sensitivity to variation management. In many contexts sources of unwanted variation, potentially detrimental to the product/system under study, are called noise factors.

It should be emphasized that the eras described above should not be taken too literally. They overlap and many interesting reliability issues are not addressed in this categorization. It may be hard to include human reliability and software reliability issues.

As noted above, there is a lack of systematic theories on design for reliability. In this collection of essays, we provide a starting point for new thinking in practical reliability improvement work. It offers the readers new principles for reliability design and tools to put these principles into action. Hopefully, it will be a thought-provoking book initiating new thinking and providing energy for further improvement of design for reliability.

---

[3] Barlow, R.E. and Proschan, F. *Mathematical Theory of Reliability*, John Wiley & Sons, Inc., New York, 1965; Barlow, R.E. and Proschan, F. *Statistical Theory of Reliability and Life Testing*. Holt, Rinehart and Winston, Inc., New York.
[4] Nowlan, F.S. and Heap, H.F. *Reliability Centred Maintenance*. Report AD/A066–579, National Technical Information Service, US Department of Commerce, Springfield, VA, 1978

We aim at readers with an interest in designing for reliability. The audience is probably advanced designers and reliability specialists. Especially, people interested in Design for Six Sigma and similar initiatives, are a target group; the techniques we promote fit well into the Design for Six Sigma framework.

Acknowledgements

The Swedish Foundation for strategic research has supported the work to develop and collect the presented material through the Gothenburg Mathematical Modelling Center.

Part of the material is previously published. Chapter 3 'Principles of robust design methodology' is published in *Quality and Reliability Engineering International* Vol. 24, 2008, published by John Wiley & Sons, Ltd. Chapter 6 'Variation mode and effect analysis: an application to fatigue life prediction' is published in *Quality and Reliability Engineering International* Vol. 25, 2009, published by John Wiley & Sons, Ltd. Chapter 9 'Model complexity versus scatter in fatigue' is published in *Fatigue and Fracture of Engineering Material and Structures* Vol. 27, 2004, published by Wiley-Blackwell. Chapter 11 'Interpretation of Dispersion Effects in a Robust Design Context' appeared in *Journal of Applied Statistics* Vol. 33, 2006, published by Routledge, Taylor & Francis Group.

We would very much like to acknowledge the cooperation and exchange of ideas with the following companies: Scania, SKF, Volvo Aero, Volvo Cars, and Volvo 3P.

Furthermore, the support from SKF to the Quality Sciences Research group is gratefully acknowledged.

About the Editors

Bo Bergman (*Division of Quality Sciences at Chalmers University of Technology*) has been SKF professor, Quality Sciences, at Chalmers University of Technology since 1999. Earlier, from 1984 to 1999, he was a professor of Quality Technology and Management at the University of Linköping. From 1969 to 1984, he was employed by Saab Aerospace in different engineering and managerial positions in the areas of reliability, quality and statistics. During that period he gained his PhD in Mathematical Statistics at Lund University and at the end of that period he was also an adjoint professor in Reliability Engineering at the Royal Institute of Technology in Stockholm. He is a member of the American Society of Quality and the Scandinavian Society of Reliability Engineers and an elected member of the International Statistics Institute. He is an academician of the International Academy for Quality

Jacques de Maré (*Department of Mathematical Sciences at Chalmers University of Technology and University of Gothenburg*) obtained his PhD in mathematical statistics in 1975 at Lund University. He worked at Umeå University from 1976 to 1979 before securing a position at Chalmers University of Technology. He became professor there in 1995. He was a visiting researcher at the University of North Carolina in 1982, at the University of California, Santa Barbara, in 1989 and at Kyushu University in Fukuoka 2005. He is a member of the International Statistical Institute and serves as a scientific advisor at the Fraunhofer–Chalmers Centre and was one of the founders of UTMIS (the Swedish Fatigue Network) and a member of the first board. He is currently working with statistical methods for material fatigue in cooperation with SP Technical Research Institute of Sweden. At Chalmers he has also worked in different ways to bring mathematical and engineering disciplines closer together.

Sara Lorén (*Fraunhofer–Chalmers Research Centre for Industrial Mathematics*) obtained her PhD in mathematical statistics in 2004 at Chalmers University of Technology: *Fatigue limit, inclusion and finite lives – a statistical point of view*. Since 2005 she has been an applied researcher at Fraunhofer–Chalmers Research Centre for Industrial Mathematics working with statistical methods for material fatigue.

Thomas Svensson (*SP Technical Research Institute of Sweden*) obtained his PhD in 1996: *Fatigue life prediction in service – a statistical approach*. He worked as a research engineer at the Technical Research Institute of Sweden (SP) from 1990 to 2001, and then at Fraunhofer – Chalmers Research Centre for Industrial Mathematics from 2001 to 2007, before returning to SP in 2007. He is a chair of the board of UTMIS (the Swedish Fatigue Network), and a member of the Editorial Board of *Fatigue and Fracture of Engineering Materials and Structures*.

Contributors

Martin Arvidsson (*Effort Consulting*) received his MSc in industrial engineering at the University of Linköping. After receiving his PhD in 2003 from the Department of Quality Sciences, Chalmers University of Technology, Martin continued to work as an Assistant Professor and Head of the same department. His PhD thesis focuses on robust design experimentation and dispersion effects in that context and today his main research interests are 'robust design methodology, reliability engineering and design of experiments'. Martin has taught many courses on reliability including a Six Sigma Black Belt course. He is now propagating the research knowledge as a consultant in industry.

Stefano Barone (*Division of Quality Sciences of Chalmers University of Technology and University of Palermo*) received his PhD in applied statistics in 2000 from the University of Naples (Italy). After having worked for two years as a post-doctoral researcher, first at the ELASIS (FIAT research centre) and then at the University of Naples and Chalmers University of Technology, he became Assistant Professor of Statistics at the University of Palermo, Faculty of Engineering. In 2005 and 2006 he served as a member of the board and then Vice President of ENBIS, the European Network for Business and Industrial Statistics. In 2008 he was a Fulbright Visiting Research Scholar at the Georgia Institute of Technology (USA), School of Industrial and Systems Engineering. Currently he is also Associate Professor of Quality Sciences at the Chalmers University of Technology and responsible for the Six Sigma Black Belt education at Master's level.

Anders Bengtsson (*Mathematical Statistics, Centre for Mathematical Sciences, Land University*) received his MSc in Engineering Physics in 2004. Since the autumn of 2004 he has been a PhD student in mathematical statistics at Lund University, Sweden. His main research interests are nonstationary processes and extreme events modelling with applications in risk and safety analysis.

Klas Bogsjö (*Scania CV AB, RTRA Load Analysis*) obtained his PhD in engineering at the Department of Mathematical Statistics, Lund University, on the topic of *Road Profile Statistics Relevant for Vehicle Fatigue*. Currently he is working at Scania.

Alexander Chakhunashvili (*Skaraborg Hospital Group*) received his MSc in management of production from the of Chalmers University of Technology in Gothenburg. After he received his PhD in 2006 at the Department of Quality Sciences, Chalmers University of Technology, he started as a Master Black Belt in healthcare. Alexander's PhD thesis deals with *Detecting, identifying and managing sources of variation in production and product development* and he has also been involved in the teaching of reliability courses, as well as the Six Sigma Black Belt course.

Ida Gremyr (*Division of Quality Sciences at Chalmers University of Technology*) received her MSc in industrial engineering and Management at Chalmers University of Technology in Gothenburg. Having completed her PhD at the Department of Quality Sciences at Chalmers University of Technologies in 2005, Ida continued to work as an Assistant Professor and master's program director at the same department. Her PhD thesis focuses on robust design methodology as a framework with a number of supportive methods. Today her main research interests are quality methods applied in product as well as process development, with a special focus on robust design methodology and Design for Six Sigma. Ida has been involved in various courses in the quality area including the Six Sigma Black Belt course.

Pär Johannesson (*Fraunhofer–Chalmers Research Centre for Industrial Mathematics*) received his MSc in computer science and engineering at Lund Institute of Technology in 1993, with a thesis within signal processing. In 1999 he obtained his PhD in engineering at mathematical satistics, Lund Institute of Technology, within the area of statistical load analysis for metal fatigue. During 2000 and 2001 he had a position as a post-doctoral researcher at the Department of Mathematical Statistics, Chalmers University of Technology within a joint project with PSA Peugeot Citroën. He spent the year 2000 at PSA in Paris, working for the Division of Automotive Research and Innovations on the topics of reliability and load analysis for fatigue of automotive components. He is currently working as an applied researcher at the Fraunhofer–Chalmers Research Centre for Industrial Mathematics in Göteborg, with industrial and research projects on load analysis, reliability, and statistical modelling, mainly with problems related to fatigue.

Per Johansson (*Volvo Powertrain AB*) received his MSc in industrial engineering from the University of Central Florida in 1998. He works at Volvo Powertrain as a quality and reliability manager. In addition, he is a PhD candidate at Chalmers University of Technology where he received a Licentiate of Engineering 2004. His main research interests are robust design methodology and reliability engineering.

Erland Johnson (*SP Technical Research Institute of Sweden*) received his MSc in 1986 in physical engineering at the University of Lund. In 1991 he received his PhD from the Department of Solid Mechanics at Lund University within the field of dynamic fracture mechanics. Erland has thereafter been working in the vehicle industry with solid mechanics and mechanical life predictions related to different failure modes for several years. In 2001 Erland became research and development manager at the Division of Mechanics at SP Technical Research Institute of Sweden. His research interests have been broadened from fracture and fatigue to also include reliability aspects. In 2009 Erland became Adjunct Professor at Shipping and Marine Technology at Chalmers University of Technology.

Åke Lönnqvist (*Volvo Car Corporation*) has an MSc engineering physics and more than 25 years experience of applied reliability engineering in automotive product development. Many years of active participation in industrial reliability organizations (e.g. member of the board in SFK-TIPS and the Scandinavian Organization of Logistics Engineers) have also provided an insight into how Reliability Engineering is applied in a wide range of industries. He is the author of several books and reports concerning applied reliability engineering. Merging and integrating reliability engineering and robustness thinking into product development processes is currently an interesting topic for the industry. This is also a main research objective for his current PhD studies at Chalmers University of Technology.

Igor Rychlik (*Department of Mathematical Sciences at Chalmers University of Technology and University of Gothenberg*) is Professor in Mathematical Statistics at Chalmers University of Technology. He earned his PhD in 1986 with a the thesis entitled *Statistical wave analysis with application to fatigue*. His main research interest is in fatigue analysis, wave climate modelling and in general engineering applications of the theory of stochastic processes, especially in safety analysis of structures interacting with the environment, for example through wind pressure, ocean waves or temperature variations. He has published more than 50 papers in international journals, is the co-author of *Probability and Risk Analysis. An Introduction for Engineers*, and visiting professor (long-term visits) at: the Department of Statistics, Colorado State University; Center for Stochastic Processes, University of North Carolina at Chapel Hill; Center for Applied Mathematics, Cornell University, Ithaca; Department of Mathematics, University of Queensland, Brisbane, Australia.

Leif Samuelsson (*Volvo Aero Corporation*) is working with method development in the fatigue area at Volvo Aero AB in Trollhättan. He received his MSc in applied mechanics at Chalmers University of Technology in 1972. He worked with structural analysis at Ericsson Radar Systems between 1972 and 1980. He proceeded with structural analysis on ships at the shipyard in Uddevalla from 1980 to 1982. In 1982 he moved to Volvo Aero, where a large part of the work was devoted to fatigue problems. He worked at SAAB Automobile from 1987 to 1992, mainly with load analysis and fatigue. In 1992 he moved back to Volvo Aero.

Part One

Methodology

Contrary to most reliability texts we promote the study of variation and uncertainty as a basis for reliability work. In Davis (2006)[1] a similar framework is described and Clausing also advocated a related approach long ago, see Clausing (2005)[2]. Robust Design Methodology (RDM) is a core technique in our approach. In most texts the RDM techniques are initiated from Failure Mode and Effects Analysis (FMEA), which was suggested in the sixties and which nowadays has widespread application in industry. As a complement, or possibly, as a replacement of FMEA we suggest an enhancement of that technique – what we call VMEA – Variation Mode and Effects Analysis. In VMEA sources of variation and uncertainties affecting important outputs are identified and assessed. In our approach, we go immediately to the root causes of failures, i.e. to sources of variation and uncertainties. When important areas for improvement are identified, we suggest the utilization of Robust Design Methodology.

---

[1] Davis, T. P. Science, engineering, and statistics. *Applied Stochastic Models in Business and Industry*, **22**: 401–430, 2006.

[2] Clausing, D. Operating window – an engineering measure for robustness. *Technometrics*, **46** (1): 25–29, 2004.

Robust Design Methodology for Reliability: Exploring the Effects of Variation and Uncertainty
edited by B. Bergman, J. de Maré, S. Lorén, T. Svensson
© 2009, John Wiley & Sons, Ltd

1

Introduction

Bo Bergman and Martin Arvidsson

1.1 Background

There is no system that always performs as intended. There are many examples of this in our daily lives; we often experience failures of automobiles, mobile telephones, computers and their software. It is also obvious in the operation of more complex systems, for example space vehicles, railway systems and nuclear power plants.

This book is about design principles and systematic methods to reduce failures and thereby improve reliability as experienced by the users of the systems. Unreliability is not only a problem for the users of the systems – the producers also suffer. Failures reduce company profitability through call-backs, warranty costs and bad will. Warranty costs alone are often estimated to be 2–15% of total sales. It is also sometimes said to be in the size of the product development costs, but that might be a problematic comparison – a small increase in product development costs to increase reliability might bring a considerable gains in reduced warranty costs. Even though warranty costs are considerable, the largest loss to a company because of low reliability is probably the loss of good will. Regardless, reliability is an important feature of current products and systems, software, hardware and combinations thereof.

Essentially, in this book we will relate to hardware systems and their reliability and how to work with reliability issues early in the product life cycle. Contrary to most reliability texts, we promote the study of variation and uncertainty as a basis for the reliability work. In Davis (2006) a similar framework is described and Clausing (2004) also puts forward a similar approach. Robust Design Methodology (RDM) plays a central role in our approach to developing high reliability hardware systems.

Even though reliability in most textbooks is given a probabilistic definition, the reliability of a product is here defined as:

'The ability of a product to provide the desired and promised function to the customer or user.'

Robust Design Methodology for Reliability: Exploring the Effects of Variation and Uncertainty
edited by B. Bergman, J. de Maré, S. Lorén, T. Svensson
© 2009, John Wiley & Sons, Ltd

Many different approaches can be used to measure this ability; most often they are related to probability theory and concern the length of failure-free operation. In most textbooks on reliability, the concept 'reliability' is in fact defined as a probability, i.e. 'the probability of failure-free operation during a certain time period and under stated conditions of usage'. However, many other measures might be relevant. In fact, as we shall see below, such alternatives might be more relevant from the point of view of engineering design.

The conventional strategy for reliability improvement work has been to utilize feedback from testing and from field usage to understand important failure mechanisms and then, in the future, try to find engineering solutions to avoid or reduce the impact from these mechanisms. Based on past experiences it has also been the practice to make predictions of future reliability performance in order to spot weak points and subsequently make improvements with respect to these weaknesses when already in early stages of the design. In his study of the room air conditioning industry, Garvin (1988) found strong evidence that the existence of a reliability group taking care of field failure experiences and giving feedback to the designers gave a positive effect on equipment reliability.

However, the conventional reliability improvement strategy has some strong limitations, as it requires feedback from usage or from expensive reliability testing. Thus, it is fully applicable only in later stages of product development when already much of the design is fixed and changes incur high costs. Consequently, there is a need for a more proactive approach. The aim of this book is to indicate some paths towards such an approach based on relations between failure occurrence, variation and uncertainty. In fact, this approach has been available for a long time, but a systematic proactive methodology has been missing. Today, we can utilize the development of RDM to make products insensitive to variation and uncertainties and thereby improve the reliability performance of products, see e.g. Davis (2006). In Chapter 6 it is argued that countermeasures to increase the reliability of systems can be divided into three categories: 'fault avoidance', 'architectural analysis' and 'estimation of basic events'. It is further argued that an essential part of these countermeasures can be realized by designing systems that do not fail despite the existence of noise factors, i.e. the creation of robust designs.

1.1.1 Reliability and Variation

Variation is everywhere – for good and for bad! Variation is at the core of life itself and to find new solutions – for creative action – we need variation in the stream of ideas and associations. However, in the manufacturing and usage of systems variation might be a problem. Understanding variation is an important aspect of management as already emphasized by Shewhart (1931, 1939) and later by Deming (1986, 1993). Especially, understanding variation is an important aspect of engineering knowledge.

In the early history of railway development failures often occurred – a derailing of a train returning from Versailles in 1842 due to a broken locomotive axle has become a well-known example of a failure due to fatigue. Similar events were systematically investigated by August Wöhler in the mid-nineteenth century. Wöhler suggested the existence of a fatigue limit and established an empirical life–load relationship, the S–N (Stress, Number of cycles failure) curve. Fatigue is a long time effect of stress variation. In later chapters of this book, see e.g. Chapters 6, 7, 9, 10 and 12, fatigue will be studied more extensively.

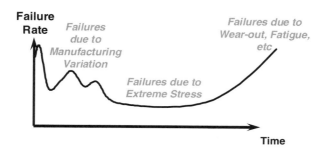

Figure 1.1 The so-called bath-tub curve, often suggested as a generic model of the failure rate of a system. The roller-coaster shape in the beginning illustrates that some of the failure mechanisms induced by the manufacturing variation might take some time (or energy) to become active.

The now well-known bath-tub curve can be used as an illustration of the relation between sources of variation and reliability, see Figure 1.1. Failures in early stages of the utilization period are usually due to variation in the manufacturing process. Due to manufacturing variation, some units are weaker than others and therefore they will fail quite early. As time passes, these weaker units will have disappeared or have been restored to a much better condition and the failure rate curve levels out. In the middle period, often called 'the best period', it is essentially only environmental high stresses that might sometimes be so severe that a unit fails. Thus, failures occur essentially independently of the age of the unit, which explains the essentially constant behaviour of the failure rate curve. At the end of the life of a surviving unit, accumulated environmental stresses and inner deteriorations make the unit weaker and more prone to fail. The failure rate curve increases again. The discussion above and Figure 1.1 relate to nonrepairable units, but similar reasoning also applies to repairable systems.

Chapter 2 gives a comprehensive overview of the initial development of reliability engineering and presents some classic reliability problems and their countermeasures. One such problem was the development of the V1 rocket during World War II. The aircraft designer Robert Lusser, one of Verner von Braun's co-workers on the unreliable V1 rockets, suggested what is sometimes called Lusser's Law: the survival probability of a (series) system is the product of the survival probabilities of the components of the system. This was not realized in the early stages of the V1 development and before Lusser joined the group. Hence the (fortunate) low survival probability.

After World War II, Lusser worked for the US military and their missile development program. He suggested a reliability design criterion based on the variation of stresses and strengths of components, see Lusser (1958). Similar types of design rules had been widely used within the mechanical industry under the name of safety margins without, however, any explicit considerations of variability. The criteria suggested by Lusser (1958) clearly take variation into account, as illustrated in Figure 1.2.

In general, the stress–strength relationship can be described as in Figure 1.3. Initially, there are varying strengths (resistances to failure) among the units produced. These units deteriorate, and as soon as a stress exceeds the strength, a failure occurs.

Chapter 4 in this book shows that the majority of failure modes identified in Failure Mode and Effects Analyses (FMEA) are actually caused by different sources of variation. Part Two of this book is basically about the relation between variation and failures.

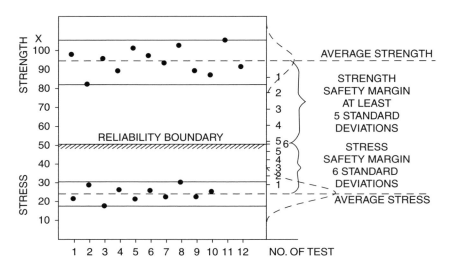

Figure 1.2   Illustration of the design criteria suggested by Lusser (1958). Lusser also related the degree of separation to the testing efforts made.

1.1.2 Sources of Variation

There is variation both in strengths and in loads. The question is what we know about them. In the RDM literature, sources of variation, also called noise factors, are often categorized as noise generated from within the system under study (also called inner noise) and as noise

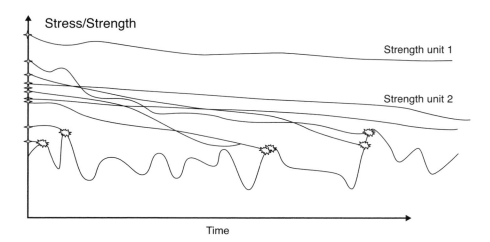

Figure 1.3   Units of different strengths deteriorate and fail as soon as a stress higher than the strength is experienced (from Bergman, 1979). It should be noted that in this figure, the strength curves of a number of units are illustrated. However, only one stress history is illustrated. In real life most units experience different stress histories.

generated from the environment or usage of the system (also called outer noise). An example of inner noise is deterioration during usage. Outer noise is illustrated by noise due to variation between customers and their use of the system and also to the environment in which the system is used. Davis (2006) instead classifies sources of variation as demand and capacity types of variation. This is a fruitful way of classification as it relates to earlier reliability work; load and strength variations can be seen as special cases of demand and capacity variations, respectively. The demand type sources of variation are:

- variation between customers (different customers/users will use the system in different ways)
- usage variation within customer (variation within and between duty cycles)
- external environmental variation (the external variation will differ between different units and also in time for the same unit)
- system environmental variation (when the unit is in fact a module in a modularized system, the demand from the system might differ depending on actual system configuration; also other types of variation in the system environment, e.g. due to deterioration of other system components, should be included here).

The sources of variation affecting capacity are:

- variation of part characteristics due to production variation
- variation of part characteristics due to usage (deterioration, fatigue, wear, corrosion, etc.). (It should be noted that here we have an interaction, e.g. between duty cycle variation and deterioration.)

In Chapter 5 we will make a more detailed analysis of different sources of variation and their impact on reliability as experienced by the customer.

Figure 1.4 displays a common view on how sources of variation influence the failure risk at a certain point in time. However, it is necessary to be cautious when viewing the influence of noises in the way indicated by the graphs in Figure 1.4. There might not be any well-defined distributions whatsoever! Many of the noises are certainly not 'under statistical

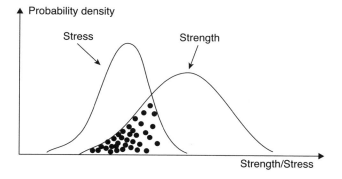

Figure 1.4   Demand/Capacity (stress/strength) distributions at a certain time.

control' according to the criteria suggested by Shewhart (1939). If they are not under statistical control we cannot utilize past history for predictions about the future.

The obvious course of action is to make the separation between demand and capacity as large as economically possible. An aid in this work could be to utilize, for whatever it is worth, past observations of demands and capacities. Chapter 7 discusses the advantages and disadvantages of two different ways of approaching this problem, the safety factor approach and the load–strength method. The apparent drawback of the experience based safety factor approach is that it cannot be updated in a rational way as its origin is difficult to analyse. In theory, the probability based load–strength approach is sound. However, as the available information about the input is often scarce and the usefulness of the result to a large extent depends upon the distributions of the input, the approach is less useful in practice. Further, Chapter 7 presents an alternative approach that combines the advantages of the two methods.

1.1.3 Sources of Uncertainties

It is possible to categorize the causes of malfunction, i.e. product inability to provide desired function, in the following way:

1. This category is related to human mistakes, such as misuse, unforeseen physical effects or unforeseen extreme events.
2. This category is lack of knowledge of identified physical behaviour, and variation in future usage (this is sometimes referred to as epistemic uncertainty)
3. There are random variations both in product features and in usage; this gives rise to uncertainties that are sometimes called aleatory.

In all these categories there is a need for a rational treatment of the uncertainties and variations in order to obtain reliable products, and in this book, for example in Chapters 6 and 7, we put forward ideas that make it possible to take advantage of statistical tools in this work.

The first category, related to human mistakes, is probably the most important cause of reliability successes or failures. However, it is difficult to describe problems from this category in mathematical terms. See also Chapter 13 where these aspects are briefly discussed. When the predominant failure modes are related to nonrandom events, it is still possible to make use of statistical tools, namely in order to make products robust against different events.

The second category can be treated with statistical tools by using the so-called Bayesian perspective: lack of knowledge may be modelled as a statistical population of unknown errors and put into the analysis by the application of statistical measures of the uncertainties from, for instance, engineering judgements. Such a methodology makes it possible to combine unknown errors with true variation effects in order to identify the most important weaknesses in the product reliability. This concept is developed in Chapter 9 of this book.

The treatment of the third category with statistical tools is straightforward and well established. Here, some important tools for treating random variables and their interactions are treated in some detail, namely the use of design of experiments for making products robust against variation.

Powerful computer simulation tools are available for treating random variables. However, there is a risk associated with the high precision probability measures that are the outcome of these advanced statistical treatments. If the uncertainties in the input variables are not taken into account, the precision in the obtained probability measures will be far less accurate than that given using the simulations. Chapter 10 uses the statistical framework for finding an optimal complexity in modelling, emphasizing the trade-off between the amount of input knowledge and the model complexity. This trade-off is treated in Chapter 9.

When failures occur, it is not always the variation itself that is the problem but rather the uncertainty about the size of the load on the equipment. Load assumptions are often based on inadequate knowledge of customer use. If misjudgments are made, seemingly harmless sources of variation might result in a failure. Thus, the probability distributions shown in Figure 1.4 should be the predictive distributions of demands and capacity, respectively. Consequently modelling uncertainties both with respect to kinds of models (e.g. linear or more complex relationships) and parameter uncertainties should also be included.

Today there is a big difference between the complexity of the models used in industry and those developed in academia. Chapter 10 deals with the question of what complexity level to use in certain situations. It is concluded that the optimal model complexity can be found by means of statistical modelling of prediction uncertainty. A particularly problematic issue is the question of future occurrences of what is called special or assignable causes of variation; i.e. if the processes underlying the demand and capacity are not under statistical control. How do we ensure that all the sources of variation and uncertainties are predictable?

1.2 Failure Mode Avoidance

1.2.1 Insensitivity to Variation – Robustness

When designing a product we want the two distributions in Figure 1.4 to be separated. Davis (2006), following recommendations made at the Xerox Company in the early eighties (see Clausing, 1994), describes his approach as *Failure Mode Avoidance* – striving to avoid failures by separating demand and capacity as much as possible. Note that we might have some problems in defining the 'distributions' as discussed above. It should also be noted that the separation of the 'distributions' in Figure 1.4 might be achieved in many different ways.

Traditionally, separation has been achieved by reducing or limiting different sources of variation (noises). Different kinds of tolerances have been imposed in manufacturing in order to achieve separation. Sometimes also limitations with respect to usage have been prescribed. Furthermore, via burn-in (or proof-testing), the left-hand tail of the capacity 'distribution' has been reduced before the units have been put into operation.

In some situations there exists an even better solution to reliability problems than separation of load and capacity. In situations when the system demand has to do with a specific noise factor, it is sensible to consider the possibility to redesign the system in such a manner that the system is made independent of that noise factor. This strategy is much in line with what Taguchi (1986) refers to as parameter design, where system prototypes are made insensitive to different kinds of noise factors. Only if the measures to be taken are too expensive or not quite sufficient, further separation is necessary. In such situations, utilizing tolerance design in the more traditional way of achieving separation is recommended. However, the literature has to a large extent focused on improving already existent system prototypes using parameter design backed up by experimental methods. Less focus has been put on the possibility of designing robust system concepts based on creativity. The basis for such a design is that a creative

Figure 1.5   The first self-aligning SKF ball bearing. (Reproduced by kind permission of the company Photo Library, SFK.)

solution is sought where some important noise factors are made irrelevant or at least much less important to the reliability of the product. Thus, the difference compared to parameter design is that a brand new design solution is developed, not an improved version of an already existing design. Below we first illustrate 'creative robust design'. This is followed by a discussion of Robust Design Methodology in terms of parameter design.

1.2.2 Creative Robust Design

Sometimes it is possible to make a change to the product so that the failure mode is avoided or at least its occurrence is made dramatically less probable. The self-aligning ball bearing (see Figure 1.5) invented by Sven Winqvist in 1907 is an example of a product with inherent robustness. Before the self-aligning ball bearing was introduced, misalignments of the shaft had a disastrous influence on bearing life. Such misalignments can for example be caused by subsidence in premises where machines are located. By using outer rings with spherical raceways, the self-aligning ball bearing was designed to allow for misalignments of the shaft. Looking for such solutions with inherent robustness is an important issue for increasing reliability.

At SKF[1] this tradition of creating products robust against sources of variation has continued. In 1919 Palmgren created a similar robust roller bearing also insensitive to shaft angle variations, and in 1995 Kellstrom created the toroidal roller bearing CARB℗, which is insensitive to both radial and axial displacements.

1.3 Robust Design

The objective of Robust Design Methodology is to create insensitivity to existing sources of variation without elimination of these sources. Thus, parts of the creative robust design discussed above are also included. It might not be possible to suggest a strict methodology

---

[1] SFK Group is the leading global supplier of products, solutions and services within rolling bearings, seals, mechationics, services and lubrication systems. Services include technical support, maintenance, condition monitoring and training.

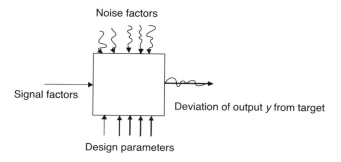

Figure 1.6   The P-diagram. (Adapted from Phadke, 1989.)

to identify creative solutions. However, some necessary prerequisite could be achieved in a systematic way. One such prerequisite is the understanding of important sources of variation when already in the very early stages of product development.

In this section we will discuss RDM to some extent. However, we will not go into depth on this as there are a lot of RDM descriptions already in many different settings (see e.g. Gremyr, 2005, and references given there). Davis (2006) also has an interesting discussion with special emphasis on reliability.

Chapter 3 discusses agreements and disagreements on the view of the robust design concept. Efforts to create robust designs have often been seen as synonymous to parameter design on already existent system concepts. Almost no focus has been put on the development of tools and aids helpful to come up with creative robust design solutions. The main conclusion drawn in Chapter 3 is that robust design is an aim that should be emphasized in all stages of design.

1.3.1 Product Modelling

In RDM the output from systems are described as functions of signal factors, design parameters (often denoted control factors) and noise factors as illustrated in the P-diagram (P stands for product/process) in Figure 1.6. These functions are commonly denoted transfer functions. The levels of the signal factors are set by the user of the system to express the intended value of the output. The levels of the design parameters can be determined by the designer of the system whereas the levels of the noise factors cannot be controlled during operating conditions. The problem focused in RDM is to identify system solutions with small output variation for different levels of the signal factors despite the existence of noise factors.

One problem in system design is that there might be a large number of outputs potentially affected by noise factors of different kinds. There is a need to give more attention to a few very important outputs and corresponding design elements. It is unwise to start working on issues of minor reliability importance as long as more important issues are not adequately handled. One way to select important areas for further analysis is to use FMEA, a well-known tool within reliability engineering. At the same time, a checklist with relevant noise factors can serve as inspiration when tracking possible causes to failure modes identified in FMEAs. Chapter 4 concludes that the majority of failure causes can be attributed to noise factors whereas a minor number of failure causes are related to the nominal performance of products.

Chapter 5 presents a systematic tool similar to those of fault tree analysis and FMEA. The tool is called Variation Mode and Effects Analysis (VMEA) and there exist a number of different variants depending on how detailed knowledge is available of the studied system. Different variants might be thought of as the different variants of FMEA. Unlike FMEA, however, this tool is top-down rather than bottom-up. When extensive knowledge is available of the system studied, it is possible to perform a variant of VMEA called probabilistic VMEA. Not only random variation but also uncertainties are considered. A case study where this kind of VMEA is applied to fatigue life prediction is presented in Chapter 6.

Once it is understood how outputs are affected by signal factors, design parameters and noise factors (as a mathematical function or as a simulation model), possible interactions between design parameters and noise factors can be utilized to reduce output variation of the system. This is achieved by appropriate settings of design parameters that make the system insensitive to noise factors.

When the transfer function is known or has been estimated by use of design of experiments, the error transmission formula and Monte Carlo simulation are often tools used to study how variation in input variables transmits to one or many output variables. However, the usefulness of these methods is largest when there is variation around the nominal values of design parameters. Such a noise factor is for example variation of part characteristics due to production variation. In Chapter 8 advantages and disadvantages of using the error transmission formula and Monte Carlo simulation are discussed. One important conclusion is that the error transmission formula is advantageous when seeking robust design solutions.

It is often favourable to conduct a computer experiment using Design of Experiment principles even though the transfer function is known in principle. The reason for this is that it is often too complicated to see through when it comes to failure mode avoidance, i.e. finding design parameter settings making the product insensitive to the noise factors. If by analytical means it is not possible to find the transfer function, physical experiments utilizing the Design of Experiment principles have to be made. The basic idea when searching robust solutions by use of Design of Experiments is to vary design parameters and noise factors in the same experiment.

An experiment of this type was conducted at Saab AB Sweden. The aim of the experiment was to improve the robustness of the manufacturing process of composite material used in military fighter airplanes. Three design parameters were identified by operating personnel and quality engineers: curing temperature, pressure and holding time. Five noise factors characterizing the incoming material were identified: proportion of hardener, thermoplastic content, proportion of epoxy, material storage time and type of process. All eight factors included in the experiment were varied at two levels each. The response variable bending strength was used for characterization of the composite material. For the sake of confidentiality, the levels are here coded -1 and $+1$.

An analysis of the experiment showed that the process pressure, the material storage time and the interaction between these two factors had a significant effect on the bending strength. The direction of this influence is shown in Figure 1.7, which shows that keeping the pressure on a high level makes the bending strength of the composite material more insensitive to the influence of the material storage time. Thus, if the pressure is kept at the high level there is no reason to control the material storage time, which means a cost saving.

In many situations it is too expensive or even impossible to vary noise factors at fixed levels in experiments. An alternative in these cases is to allow noise factors to vary randomly during

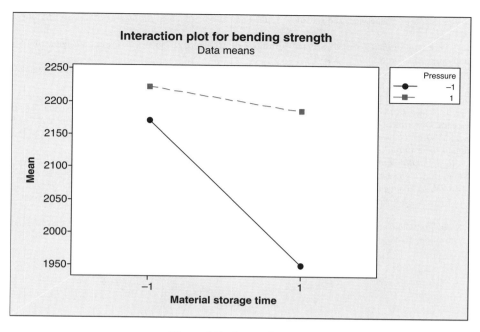

Figure 1.7 Interaction plot.

the experiment and to capture the influence of noise factors via replicates of experiments. Since replication of experiments is often not an alternative for economic reasons, the influence of noise factors that vary randomly has to be captured by use of residual based dispersion analysis. In Chapter 11 it is shown how interactions between design parameters and noise factors that vary randomly during the experiment are manifested as dispersion effects of the design parameters.

1.4 Comments and Suggestions for Further Reading

For early discussions on the economic impact of unreliability, see e.g. Garvin (1988). For some more recent references, see e.g. Johnson and Nilsson (2002) and Curkovic et al. (1999), who show a strong relationship between reliability of products and business performance in the automotive industry.

References

Bergman, B. *Design of Reliable Products, Part I, Introduction and General Principles*. Saab Scania AB, Aerospace Division, FTK 81:08, 1979. (In Swedish).

Clausing, D. *Total Quality Development – A Step-by-Step Guide to World-Class Concurrent Engineering*. ASME Press, New York, 1994.

Clausing, D. Operating window – an engineering measure for robustness. *Technometrics*, **46**(1): 25–29, 2004.

Curkovic, S., Vickery, S. K. and Droge, C. Quality and business performance: An empirical study of first tier automotive suppliers. *Quality Management Journal*, **6**(2): 29–40, 1999.

Davis, T. P. Science, engineering, and statistics. *Applied Stochastic Models in Business and Industry*, **22**: 401–430, 2006.

Deming, W. E. *Out of the Crisis*. Massachusetts Institute of Technology, Center for Advanced Engineering Study, Cambridge, MA, 1986.

Deming, W. E. *The New Economics for Industry, Government, Education*. Massachusetts Institute of Technology, Center for Advanced Engineering Study, Cambridge, MA, 1993.

Garvin, D. A. *Managing Quality, The Strategic and Competitive Edge*. Free Press, New York, 1988.

Gremyr, I. *Robust Design Methodology – A framework and supportive methods*. PhD thesis, Division of Quality Sciences, Department of Technology Management and Economics, Chalmers University of Technology, 2005.

Johnson, M. D. and Nilsson, L. The importance of reliability and customization from goods to services. *The Quality Management Journal*, **10**: 8–20, 2002.

Lusser, R. *Reliability through Saftey Marignals*. U.S. Army Ordnance Missile Command, 1958.

Phadke, M. S. *Quality Engineering using Robust Design*. Prentice Hall, Englewood Cliffs, NJ, 1989.

Shewhart, W. A. *Economic Control of Quality of Manufactured Products*. Van Nostrand, New York, 1931.

Shewhart, W. A. *Statistical Method from the Viewpoint of Quality Control*. Dover Publications, Washington, DC, 1939.

Taguchi, G. *Introduction to Quality Engineering – Designing Quality into Products and Processes*. Asian Productivity Organization, Tokyo, 1986.

2

Evolution of Reliability Thinking – Countermeasures for Some Technical Issues

Åke Lönnqvist

2.1 Introduction

If the saying 'history repeats itself' is true for reliability engineering, it is obvious that a good knowledge of the historical course of events would be an advantage for assessing the direction on future reliability development.

'The farther backward you can look, the farther forward you are likely to see.'[1]

The purpose of studying drivers of early reliability engineering deveopment is to create a basis for better understanding of how current reliability engineering and methodologies can develop further. There are many historical examples, some of them discussed below, that show how technical problems have initiated improvement activities, both purely technical improvements and also development of standards, methods and models for reliability analysis and assessment.

A general question is 'what can we learn from past experiences of solving problems for future development of reliability engineering'? Are there historical approaches that can be used for finding countermeasures which also could be applied today? Are some of the approaches from the past not relevant anymore? With this historical perspective the research question is here formulated as:

---

[1] Quote attributed to Winston Churchill (Saleh and Marais, 2006).

Robust Design Methodology for Reliability: Exploring the Effects of Variation and Uncertainty
edited by B. Bergman, J. de Maré, S. Lorén, T. Svensson
© 2009, John Wiley & Sons, Ltd

Are the types of approaches for reliability engineering countermeasures, prompted by historical technical concerns, still valid as motivators for similar reliability activities in current product development?

This is an extensive issue and to give an exhaustive answer to the questions above is far beyond the scope of this chapter, but by studying some of the historical examples and the reliability engineering countermeasures taken, we can learn something for the development and application of future reliability methods and countermeasures. It is the belief of the author that a discussion of past experiences can create ideas for future deployment of methods and approaches for reliability assurance in product development.

A central element in reliability engineering is the definition of 'failure'; activities during product development aim to avoid failures and eliminate or minimize their consequences if they still could occur. Avoiding 'failures' is also a central concern in other methodologies. An example is Off-Line Quality Control (Taguchi and Wu, 1979), here referred to as robust engineering. This is a methodology that focuses on reducing the probability of failure by identifying and handling sources of variation and making the product less sensitive to the factors (called noise factors) that are the causes of failures. An interesting question is whether this is a new approach for reliability engineering?

Here, countermeasures to technical problems related to failures will be discussed. How to classify the countermeasures has been one concern in the study. It was found that the International Standard *'Dependability management, part 3-1 Application guide-analysis techniques for dependability – Guide on methodology'* (IEC60300-3-1, 2003) offered an interesting categorization of dependability analysis methods, with regard to their main purpose. Since this is an established and accepted way of classifying methods, it was decided that this categorization should be used in a more general way, not only for methods but also for other identified countermeasures.

The first category is for methods that are aimed at 'avoiding failures'. Examples of such methods are part derating, and selecting, and stress–strength analysis. The second category consists of methods for studying the architectural structure of a design and allocating dependability. This category is divided into one section with so-called 'bottom-up' methods that deal with single faults, and another section with 'top-down' methods that are able to account for effects from combinations of faults. Examples of 'bottom-up' methods are Event Tree Analysis (ETA), Failure Mode and Effect Analysis (FMEA) and Hazard and Operability Study (HAZOP). Examples of 'top-down' methods are Fault Tree Analysis (FTA), Markov Analysis, Petri Net Analysis and Reliability Block Diagram (RBD). The third category consists of methods for estimation of measures for basic events. Typical methods in this category are Failure Rate Prediction, Human Reliability Analysis (HRA), Statistical Reliability Methods and Software Reliability Engineering (SRE).

These three categories represent well-established approaches to handling reliability issues in product and process development, and for that reason the categories are used throughout this chapter as a basis for further discussion on reliability engineering and associated countermeasures, although they are given a somewhat wider interpretation as can be seen below.

1. 'Fault avoidance' consists of activities that aim to avoid the occurrence of failures. This may include 'derating' of components and activities to make the design robust against factors affecting the reliability.

2. 'Architectural analysis' is here defined as activities aimed to influence the structural design of a product with the purpose of making it more reliable, robust and safe. This means that both the probability of occurrence of a failure and the consequences of a failure are taken into consideration.
3. 'Estimation of basic events' includes activities that aim to give quantitative measures for the studied events or failure mechanisms.

This study presents some examples of historical technical reliability concerns and related countermeasures taken to improve the situation. The discussion on countermeasures and their influence on reliability engineering approaches will utilize the three categories above to classify the countermeasures.

2.2 Method

Literature on reliability engineering history seems unanimous in the opinion that its origin is to be found in the rapid technical development around World War II, (e.g. Bergman and Klefsjö, 2001; Green and Bourne, 1972; Saleh and Marais, 2006). Studies of some important technical steps, described in literature on the technical development around World War II, gave an overall picture of technical and reliability engineering development. This picture was complemented with interviews with persons involved in the development of reliability engineering from the mid 1950s, in order to guide the selection of cases; 'interview' sources are indicated in the text. Only well-known technical concerns with clear connections to reliability were selected, but they probably represent only a fraction of all cases possible to select. The intention was to select a number of major concerns that have influenced the way of thinking and reacting from a reliability engineering point of view, and which are consequently interesting for this study. This article is not claimed to be exhaustive and it is acknowledged that there is a subjective component in the selection of cases.

The Science Citation Index (SCI) was used for searching relevant literature on reliability and technical development. Some references on technical issues around World War II were found in special literature on military history (SMB – Svenskt Militärhistoriskt Bibliotek). International reliability standards (IEC – International Electrotechnical Commission) were used as references for methods for reliability engineering, and interviews with persons participating in the development of these standards and Swedish industrial reliability practices were used as additional, and important, sources of information.

2.3 An Overview of the Initial Development of Reliability Engineering

Before describing some examples of technical concerns that have influenced the development of reliability engineering, it may be useful to present a brief overview of the history of reliability engineering. As already mentioned, the origin of reliability engineering is often claimed to be the very rapid technical development around World War II (e.g. Bergman and Klefsjö, 2001; Green and Bourne, 1972; Saleh and Marais, 2006). Experiences of technical problems in many areas had created a latent concern that only waited to be resolved. The trigger for reliability engineering, as a separate discipline, is often considered to be the problems associated with the vacuum tube, replaced five times as frequently as other equipment during World War II. This concern initiated a series of military studies and ad hoc groups and the establishment of the Advisory Group on Reliability of Electronic Equipment (AGREE) in 1952, sometimes regarded as the true advent of reliability engineering (Saleh and Marais, 2006).

Prior to World War II the concept of reliability had more or less been intuitive, subjective and qualitative (Dhillon, 1986). For example, the fact that an aircraft engine might fail initiated qualitative comparisons between one and two-engined aircraft and between two and four-engined aircraft (Green and Bourne, 1972).

Important for the emerging discipline of reliability engineering were the development of mathematical statistics and the increasing use of mass production, popularized by Henry Ford around 1910. Walter A Shewhart's statistical approach to process control, developed in the 1920s and 1930s, made it possible for economic control of quality in mass production (Saleh and Marais, 2006) and may have inspired a statistical approach to some reliability issues.

Reliability efforts developed into three areas. One was to collect field data, determine technical causes and use the information for technical improvements. The second development that occurred was that quantitative reliability requirements were introduced in specifications, which initiated contractual aspects on reliability. Given specifications on reliability there was a need for the third area of development, some kind of formalized way of measuring or estimating product reliability before components and systems were built and tested. This led to the development of models for reliability prediction, a technique that provides models for calculating/estimating components and systems failure rates. The first reliability predictions were probably made by the Germans in 1942, during the V1 missile development, but now the method has become more formalized.

During the late 1950s and the 1960s, reliability engineering proceeded along two major pathways. One was an increased specialization in the discipline with improved statistical techniques, studies of reliability physics and structural reliability (Saleh and Marais, 2006). Studies of environmental factors that influenced the reliability of components were an important part of specifications and testing procedures (*interviews*), an approach not so different from some of the efforts in robust engineering. A second trend was that the focus shifted from studying component reliability to the study of system reliability (*interviews*), a natural development when systems became more complex and component interaction became more important.

It was also during the 1960s that the currently used definition of dependability, with the associated attributes, reliability performance, maintainability performance and maintenance support performance (Figure 2.1), was established in international reliability standards (IEC).

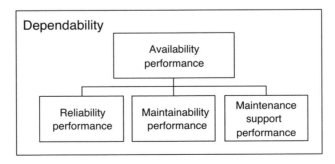

Figure 2.1 Relation between availability, reliability, maintainability and maintenance support performance according to IEC international standard.

The further development, from the 1970s and onwards, was affected by concerns over system complexity, safety aspects, human reliability and software (Saleh and Marais, 2006).

2.4 Examples of Technical Issues and Reliability Countermeasures

2.4.1 Severe Consequences

Many problems with severe consequences occur early in industrial development as well as later on, due purely to fatigue problems with components and equipment. Fatigue was, for example, a major concern during the development of the railway. Below are some other examples of physical fatigue problems that have influenced reliability engineering.

A fatigue problem in early industrial development of a chain hoisting rope (Bergman, 1981) that caused frequent fractures due to bending stresses was solved by changing technology. The iron chain was replaced with a stranded wire rope. From a reliability point of view it can be seen as the serial structure (chain) being replaced by a parallel structure (stranded wire) that drastically reduced the risk of a sudden complete fracture by introducing a kind of redundancy and a more fail-safe solution.

By studying and understanding the physical phenomena of fatigue that caused problems in many industrial applications, it was possible to control and predict the crack growth and thus plan and conduct appropriate preventive maintenance for components and systems. Examples of tools developed and used for knowledge of the physical mechanisms were Wöhler diagrams, Palmgren–Miner's law and Weibull plots.

The aviation industry was faced with several severe fatigue problems (1930–40) when they started to use steel bodies for aircraft (Wanhill, 2002). From the early 1950s, the fatigue design principle was the *safe-life* concept, which stated that the entire structure was designed to achieve a satisfactory fatigue life with no significant damage.

The DeHavilland Comet crashes in 1954 and other experiences showed that cracks could sometimes occur much earlier than anticipated, owing to limitations in the fatigue analyses, and that safety could not be guaranteed on a *safe-life* basis without imposing uneconomically short service lives on major components of the structure.

These problems were addressed by the adoption of the *fail-safe* design principles in the late 1950s. In a *fail-safe* approach the structure is designed to achieve a satisfactory life with no significant damage. However, the structure is *also* designed to be inspectable in service and able to sustain significant and easily detectable damage before safety is compromised. These latter requirements were met mainly by employing structural design concepts having multiple load paths, with established residual strength requirements in the event of failure of one structural element or an obvious partial failure, a kind of *redundancy*.

The F-111 is an unusual aircraft: it is a variable geometry 'swing-wing' fighter–bomber, and it uses high-strength steel in major airframe components. On 22 December 1969, a F-111 lost the left wing during a training flight. An immediate on-site investigation revealed a flaw in the lower plate of the left-hand wing pivot fitting. This flaw had developed during manufacture and remained undetected despite its considerable size: 23.4 mm × 5.9 mm. A limited amount of fatigue crack growth occurred in service before overload fracture of the plate, which resulted in the immediate loss of the wing.

After much research, the US Air Force provided and mandated new guidelines to ensure aircraft structural integrity. These guidelines became known as the *damage tolerance* philosophy. This approach differs from the original *fail-safe* design principles, developed for commercial transport aircraft after the Comet crashes, in two major respects:

1. The possibility of cracks or flaws in a new structure must be taken into consideration.
2. Structures may be inspectable or noninspectable in service, i.e. there is an option for designing structures that are not intended to be inspected during the service life:
   - inspectable structures can be fail-safe or slow-flaw growth structures
   - noninspectable structures may still be classified as damage tolerant provided they can qualify as slow-flaw growth, which in this case means that initial damage must not grow to a size causing failure during the design service life.

Part of the solution was then to make the design *resistant and insensitive* to manufacturing imperfections.

The main focus on safety studies seems to have shifted due to what technical issues were of current interest. Below are examples of architectural design of aircraft, missiles and safety analyses related to space development.

In the expansion of the aircraft industry after World War I, the fact that an aircraft engine might fail was partly instrumental in the development of multi-engine aircraft (Green and Bourne, 1972). At that time, purely qualitative comparisons were made between one- and two-engined aircraft and between two- and four-engined aircraft. As the number of aircraft increased, more information on aircraft system failures for a given number of aircrafts over a given length of time was collected. This led in the 1930s to expressing reliability, or unreliability, as the *average number of failures* or a *mean failure-rate*. Given data like these, reliability criteria for aircraft were discussed. Safety levels began to be expressed in terms of maximum acceptable failure rates. In the 1940s, requirements were formulated as follows: 'the accident rate should not exceed, on average, 1 per 100,000 hours of flying time'.

During the development of the French–British Concorde aircraft (Bergman and Klefsjö, 2001) specifications were set in terms of probabilities. Potential failures were classified as less severe, severe, critical and catastrophic. For example, the probability for a catastrophic failure, a crash, was allowed to be maximally 10^{-7} per flight hour.

Systems architecture was a concern during the German development of the V1 and V2 missiles in World War II. The development of the V1 started in 1942, based on the concept that a chain is not stronger than its weakest link. But the first series of V1s experienced 100% failures. The mathematician Robert Lusser was then called in and with a calculus of probability it was realized that a large number of fairly strong 'links' can be more unreliable than a single 'weak link', an approach today known as Lusser's law. Further development based on this idea gave rise to considerable improvement in the reliability of the V1; 60% of all flights were successful and similar figures also applied to the V2 missile. It is interesting to see that the experiences from unmanned earth satellites in 1972 suggest that the reliability of such devices was still in the region of 60–70% successful flights (Green and Bourne, 1972). Here it appears that although there has been a basic improvement in the component part reliability, this has been offset by the increase in *system complexity*.

At the onset of the Apollo programme, NASA risk analyses were generally accepted as the basis for technical decisions. But failure frequencies were estimated with conservative values to account for uncertainties, and it became obvious that this conservative approach gave too high estimates of the failure frequencies, contradicted by facts that had a negative influence on the acceptance of the quantitative risk analysis method itself.

In January 1986, the Challenger space shuttle had an accident during flight 51L (Paté-Cornell and Dillon, 2001). Prior to that accident, NASA had mainly performed qualitative analyses, e.g. FMEAs, and created a Critical Items List to identify a system's weak points and manage potential failure risks accordingly. In addition to this qualitative analysis, some quantitative risk analyses had been performed, but these were highly constrained and limited by the definition of their scope and restrictions on data sources and had to rely on the opinions of NASA experts as reliability data. After the Challenger accident it was concluded, e.g. by the Rogers Commission which investigated the accident, that the perception of the risks had been too optimistic and also that the risk analysis methods needed to be improved. Therefore, in the late 1980s and early 1990s, probabilistic risk analysis was again judged to be a better approach than qualitative risk analysis. In the following years of the 1990s, a number of pilot studies were undertaken. NASA also commissioned two 'proof-of concept' studies that revealed the strengths of the probabilistic approach to identify potential causes, especially for more complex events, and the approach was included as decision support.

Reliability countermeasures against the severe consequences mentioned above can be summarized as:

- Category 1 'fault avoidance'
  - introducing 'damage tolerance' as a means of being less sensitive to manufacturing variations (robustness)
- Category 2 'architectural analysis'
  - introduce redundancy in the chain example
  - redundancy in the *fail-safe* concept
  - number of aircraft engines
  - probabilistic approach to V1 and space programme system design
- Category 3 'estimation of basic events'
  - developing Wöhler diagrams, Palmgren–Miner's law and Weibull plots for better understanding and handling of the fatigue phenomena and calculation of expected fatigue life
  - requirements on probability for specified consequences of failures for aircraft
  - quantified risk analysis for Space Shuttles.

2.4.2 Defective Component

During World War II, electronic tubes were replaced five times more frequently than other equipment (Saleh and Marais, 2006). This observation prompted several investigations and ad hoc groups as mentioned earlier. According to Denson (1998) one of the groups in the early 1950s formulated the improvements required as follows:

- better reliability-data collection from the field
- better components need to be developed
- quantitative reliability requirements need to be established
- reliability needs to be verified by test before full-scale production
- a permanent committee needs to be established to guide the reliability discipline.

The last proposed improvement was implemented in the form of the Advisory Group on Reliability of Electronic Equipment, which is often regarded as the advent of reliability engineering.

There were different countermeasures. One was to find and correct root causes of the field problems experienced. Another was the specification of requirements, which in turn led to the desire to have a means of estimating the reliability performance before the equipment was built and tested so that the probability of achieving the reliability goal could be estimated. This was the beginning of reliability prediction, traced back to the document 'Reliability Stress Analysis for Electronic Equipment' released by the Reliability Analysis Center (RCA) in November 1956, which included the concept of activation energy and the Arrhenius relationship for modelling failure rates (Pecht and Nash, 1994).

When the new technology, transistors, was developed it enabled completely new architectures for electronics. An example of the technical challenge with vacuum tubes was the first electronic computer, ENIAC (Electronic Numerical Integrator and Calculator), operating between 1945 and 1955; in the beginning it was used for complex nuclear calculations during the development of the atomic bomb. The ENIAC contained nearly 18,000 vacuum tubes at a time when many experts thought that an electronic device should not contain more than 300 tubes (Barkley, 1994). Although attention to reliable operation had been given in the system design of ENIAC, the result was an availability of 50–60%. Since around 90% of the service interruptions in 1951 were caused by vacuum-tube failures, the maintenance personnel had to test incoming vacuum tubes, prepare and test plug-in circuits, and locate the causes of malfunction. Apart from the design, the availability achieved was due to a systemized tube-surveillance programme and the diligence of an effective maintenance organization.

Reliability countermeasures taken in the situations mentioned above can be categorized as follows:

- Category 1 'fault avoidance'
  - field follow-up of equipment to gather reliability data on root causes
  - efficient preventive maintenance (ENIAC)
- Category 2 'architectural analysis'
  - new technology (transistors)
  - product design for maintainability (ENIAC)
- Category 3 'estimation of basic events'
  - reliability predictions, initially for electronic equipment
  - physics of failure mechanisms and their modelling, the Arrhenius equation for temperature dependence being one example.

2.4.3 Undesired Production Variation

The concept of having guns with interchangeable parts, an idea from the French gunsmith Honoré Blanc, was developed by Eli Whitney in the USA in the late 1790s (Saleh and Marais, 2006). As a solution to the need for increasing production capacity and improving the ability for maintenance in the field, the arms industry had realized that parts should not be produced and assembled as 'individuals' but needed to be standardized. For example, during the American civil war, the Springfield armoury produced over 300,000 rifles for the union army in 1863, close to the average production of the Ford Model T, which is traditionally associated with the start of mass production. The production of the highly standardized Ford Model T started in 1908 with a production of 6000 vehicles, which increased rapidly to 800,000 in 1919. From 1910 production took place in the specialized Highland Park plant. The mass production was based on standardized exchangeable parts ('armoury practice'), detailed knowledge of the production process, specialized production equipment, and process flow oriented production with centrally controlled speed of the assembly lines. The exchangeable parts reduced the need of fitting operations by skilled workers and the working operations were changed into frequently repeated short moments in the assembly. This created a changed structure of the work force with less skilled workers and an increased share of the work force being supervisors, from only 2% of the work force in the workshops in Detroit 1893 to 28% production supervisors in the Highland Park factory 1917, some of them working with quality control (Berggren, 1990).

During the 1920s and 1930s Walter Shewhart developed the Statistical Process Control (SPC) method for economic steering and control of a production process. His major work, entitled *Economic Control of Quality of Manufactured Product*, was published in 1931. The statistically based method was an approach to minimize the production variation by distinguishing between assignable and chance causes. Corrective actions are taken to assignable causes whereas chance causes are seen as normal for the process and should not initiate actions.

It was not until after World War II that the method was implemented in manufacturing. Today this method has been adopted by many industries, e.g. the automotive industry, which utilizes SPC for process control and the associated capability study for evaluation of the capability of a machine or process to produce within specification limits. Requirements on capability studies and process FMEAs are included in the working procedures of the automotive industries (e.g. the Advanced Product Quality Planning (APQP) of the Automotive Industry Action Group (AIAG)).

Reliability countermeasures in the situations mentioned above can be categorized as follows:

- Category 1 'fault avoidance'
  - production inspection
  - statistical process control (SPC) flawless production
- Category 2 'architectural analysis'
  - standardization of the design
  - methods for evaluating the design of a process, e.g. Process FMEA
- Category 3 'estimation of basic events'
  - capability studies for estimating the possibility of production equipment to produce within specification limits.

2.4.4 Sensitivity to Noise Factors

A well-known example of reliability and robustness is provided by the so-called 'tile experiment' at the Ina Tile Company in 1953 (Taguchi and Wu, 1979). Difficulties of manufacturing tiles of a uniform size were caused by an uneven temperature distribution in the tunnel kiln. Proposals to improve the temperature distribution, which was the root cause, or to introduce some kind of production screening to reject tiles outside specified dimensions were all declined because the cost would be too high. After performing a small set of well-planned experiments another solution to the problem was found. Instead of costly changes to the tunnel kiln or the process, the solution was to make the process less sensitive, i.e. more robust, to the existing temperature distribution. The uneven temperature in the kiln was in other words treated as a disturbance, or a noise factor, for the process. The experiments showed that robustness could be achieved by changing the chemical compound in the clay used to produce the tiles. By increasing the lime content of the clay it was possible to considerably reduce the variation of the tile dimensions.

A quite different example of robustness was found in a textbook by G. Dahllöf (2001) on the development of radar (radio detecting and ranging) technologies during World War II. The development of the radar began in 1904 when C. Hülsmeyer patented a method to 'inform an observer of the presence of distant metallic objects'. Further developed, the radar came to play an important role in the outcome of World War II, e.g. in the Battle of Britain and against submarines in the Atlantic. Each time one of the contending parties made technology improvements to its radar system, the enemy immediately tried to find means to neutralize the technology. One of the countermeasures, used by the British, was to disturb the enemy radar with an artificial 'noise factor'. During an air attack aircrafts dropped thousands of small aluminum strips, called 'window', a technique used for example during the heavy air raids against Hamburg. The 'noise factor' more or less blocked the whole German radar system with echoes and made it impossible to guide the anti-aircraft defence. But countermeasures against the 'window' effect were developed. The German SN-2-radar, developed in the beginning of 1944, included several improvements and was, for example, not affected by the 'window' effect, i.e. was insensitive to the airficial noise factor. The SN-2-radar was one of the reasons for the heavy British losses of bombers over Nürnberg in March 1944. But the German success was short-lived, a few months later the British had developed countermeasures to neutralize the SN-2-radar.

Countermeasures taken for reliability improvement in the situations mentioned above can be categorized follows:

- Category 1 'fault avoidance'
  - process insensitive to noise factors (tile experiment)
  - develop technology that is insensitive to 'noise factors' (radar).

2.5 Summary

Some examples of reliability engineering response to technical problems are given above. The responses are divided into three categories described in the introduction, and a summary is provided in Table 2.1. The first comment on Table 2.1 is that although the examples do not represent all technical concerns over the years, for these technical issues two fields are empty. This is because for robustness concerns the focus is on reducing the risk of failures, by making the designs less sensitive to noise factors, and less attention is paid to the architectural design and to estimating the occurrence of failures. For the examples of safety concerns considered the focus is on the architectural design of systems and the probability of the events, and also on fault avoidance. Thus safety issues are represented in all categories.

A common feature in the 'fault avoidance' category is that several of the countermeasures are aimed at making the product more insensitive to factors influencing its ability to operate, but there are also other initiatives such as well-planned maintenance activities and reactions to process changes (Statistical Process Control) before a failure will occur. In the 'architectural

Table 2.1 Overview of reliability engineering countermeasures for technical challenges

Technical concern	Countermeasure category		
	Fault avoidance	Architectural analysis	Estimation of basic events
Severe consequences	Damage tolerance concept (resistant and insensitive)	Alternative technology, hoisting rope (redundancy) Fail-safe concept (redundancy) Number of aircraft engines (redundancy) Probabilistic approach to V1 and Space programme system design	Wöhler diagrams Palmgren–Miner's law Weibull plots Requirements on probability for specified consequences of failures for aircraft Quantified risk analysis for Space shuttles
Defective components	Field failure information system Effective preventive maintenance (ENIAC)	Alternative technology, vacuum tubes replaced by semiconductors Product designed for maintainability (ENIAC)	Reliability predictions (electronic equipment) Physics of failure, mechanisms and modelling, e.g. Arrhenius equation
Undesired production variation	Production inspection Statistical Process Control (SPC)	Standardization of systems and parts Process FMEA	Capability studies
Sensitivity to noise factors	Process insensitive to 'noise factors' (tile experiment) Technology that is insensitive to 'noise factors' (radar)		

analysis', category key issues for safety/fatigue related issues are creation of redundancy, decisions of which technology to select, standardization of parts and the product's structure for reliable and effective maintenance and producability. The category of 'estimation of basic events' includes many activities for understanding the failure mechanisms in order to estimate their occurrence, set requirements and make assessments of expected failure frequency and, for processes, the evaluation of their capacity of production within specification limits.

2.6 Discussion and Future Research

Are the countermeasures given in the examples still valid as motivators for similar reliability activities in current product development? Assessing the countermeasures in a general way it is difficults to see that the way of handling fatigue problems, for example, should not still be valid. Given a similar severe technical problem it seems reasonable to assume that an approach similar to the 'safe-life' $\rightarrow$ 'fail-safe' + 'damage tolerance' concept to ensure that consequences of a failure are not severe and that occurrences of such failures are eliminated or diminished would be applicable. Similar reasoning is valid for the other countermeasures. Thus the examples can provide a 'menu', with three categories, of ways to react to technical reliability problems.

It is interesting to note that for most technical concerns, countermeasures are taken within all categories, except for robustness. Is it so that there are other examples where this is true, or are the countermeasures taken in the other categories enough, or will robust engineering/reliability engineering develop countermeasures in the near future that fit into the 'architectural analysis' and the 'estimation of basic events' categories? Attempts in this direction may be the development of methodologies for robustness, e.g. Robust Design Methodology proposed by Gremyr (2005), which include a broader view of activities than those described by G. Taguchi in his Off-Line Quality Control concept (Taguchi and Wu, 1979), for assuring product robustness. Variation Mode and Effect Analysis (VMEA; Chakhunashvili et al., 2004) is an example of a specific, recently developed, method to evaluate what noise factors are most important for a system design. This may also be an example of robustness methods aiming at studying the 'architectural design' of a system. From the examples presented it is clear that concerns for robustness have already been an issue in the early development of reliability engineering, consequently reliability engineering and robust engineering have much in common when it comes to avoiding failure, and merging some of the approaches may create synergy effects.

The example of risk analysis for the Space programmes illustrates two important general aspects of reliability analysis. The first one is the importance of having relevant reliability methods that provide engineering with 'reasonably' correct results, so that the engineers feel confident and rely on the results to the degree that they are motivated to take proactive action, i.e. initiate countermeasures against potential problems although they have not occurred. The other important aspect is that quantitative analysis is expected to be really useful to guide engineering reliability efforts and resources to the key areas for reliability and safety assurance.

The importance of quantitative risk analysis for the future is emphasized by the historical trend from components to system-level analysis and the current technical trend towards more complex systems. If the discussion historically has been on component interactions, it may well now be on systems interaction. This is an area that may be worthy of further study. The categorization of countermeasures into the three categories, 'fault avoidance, architectural analysis, estimation of basic events', may provide a structure and a platform for discussing countermeasures against technical problems.

References

Barkley, F. W. ENIAC – A problem solver. *IEEE Annals of the History of Computing*, **16**(1), 25–45, 1994.

Bergman, B. *Fatigue from Reliability Point of View*. Saab-Scania AB, 1981.

Bergman, B. and Klefsjö, B. *Kvalitet från behov till användning*, 3rd edn. Studentlitteratur, Lund, 2001.

Berggren, C. *Det nya bilarbetet*. Studentlitteratur, Lund, 1990.

Chakhunashvili, A., Johansson, P. and Bergman, B. Variation mode and effect analysis. *Annual Reliability and Maintainability Symposium, Los Angeles*, 2004.

Dahllöf, G. *Teknikkriget som förändrade världen*. SMB, Luleå, 2001.

Denson, W. The history of reliability prediction. *IEEE Transactions on Reliability*, **47**(3): 321–328, 1998.

Dhillon, B. S. Engineering reliability management. *IEEE Journal on Selected Areas in Communications*, **4**: 1015–1020, 1986.

Green, A. E. and Bourne, A. J. *Reliability Technology*. John Wiley & Sons, Ltd, London, 1972.

Gremyr, I. *Robust Design Methodology – A Framework and Supportive Methods*. Division of Quality Sciences, Chalmers University of Technology, Goteborg, 2005.

IEC60300-3-1. *International Standard IEC 60300-3-1, Dependability Management Part 3-1: Application Guide Analysis Techniques for Dependability Guide on Methodology*, 2nd edn, 01, International Electrotechnical Commission, Geneva, 2003.

Paté-Cornell, E. and Dillon, R. Probabilistic risk analysis for the NASA space shuttle: a brief history and current work. *Reliability Engineering and System Safety*, **74**: 345–352, 2001.

Pecht, M. G. and Nash, F. R. Predicting the reliability of electronic equipment. *Proceedings of the IEEE*, **82**: 992–1004, 1994.

Saleh, J. H. and Marais, K. Highlights from the early (and pre-) history of reliability engineering. *Reliability Engineering and System Safety*, **91**: 249–256, 2006.

Taguchi, G. and Wu, Y. *Introduction to Off-Line Quality Control*. Central Japan Quality Control Association, Nagoya, 1979.

Wanhill, R. J. H. *Milestone Case Histories in Aircraft Structural Integrity*. NLR-TP-2002-521, National Aerospace Laboratory, Amsterdam, 2002.

3

Principles of Robust Design Methodology

Martin Arvidsson and Ida Gremyr

3.1 Introduction

A great deal has happened since Nair (1992) published a panel discussion on Taguchi's parameter design and robust design. Robust design in this sense refers to designing products so that they are insensitive to variation. Robinson et al. (2003) give an update on developments since 1992, concentrating on the use of statistical methods to achieve robust designs. An overview of different methods to achieve robust designs is also provided by Park et al. (2006). However, most papers on robust design do not have their main focus on the concept of robust design and its underlying principles. The purpose of this work is to investigate the principles of robust design in a literature review; principles that are not associated with specific methods but capture an underlying way of thinking.

The importance of reducing variation in product characteristics was discussed early in Japan. The ideas of Japanese engineer Genichi Taguchi were already known in his own country in the 1940s (Gunter, 1987), although they did not receive international attention until the 1980s when they were published in English. In 1979 Taguchi's ideas were translated in Taguchi and Wu (1979), a publication that was followed by many others on the subject, see e.g. Taguchi (1986), Phadke (1989), Taguchi et al. (2000). Taguchi (1986) proposes a three-step strategy for the development of products. The steps are system design, parameter design and tolerance design, with an emphasis on the use of experimental methods in the latter two steps.

This chapter is based on the article by Martin Arvidsson and Ida Gedemyr, 'Principles of robust design methodology', *Quality and Reliability Engineering International*, Vol. 24 (1), 2008, pp. 23–35, and is reproduced by the permission of the publisher John Wiley & Sons, Ltd.

The papers by Hunter (1985) and Kackar (1985) explained the essence of Taguchi's ideas in an understandable and comprehensive manner. In the late 1980s and early 1990s there was a discussion on the appropriateness of the statistical methods proposed by Taguchi. Publications contributing to this discussion were León et al. (1987), Box et al. (1988), Shainin and Shainin (1988), Welch et al. (1990), Shoemaker et al. (1991), Box and Jones (1992), Nair (1992) Lucas (1994) and Grize (1995). There are fewer publications that focus mainly on nonstatistical issues such as principles, procedures and objectives.

This chapter starts with a description of the literature search. An analysis section deals with terminology, views of variation, various procedures that can be used to achieve robust design, the objective of robust design, methods and methodologies, and the experimental approaches that are advocated. A discussion of the analysis of literature reviewed and ideas for future research are then presented. The concluding section contains a number of principles identified in a synthesis of the analysis and provides a definition of robust design methodology.

3.2 Method

The literature has been identified through searches in Dissertation Abstracts and in three databases, Compendex, Science Citation Index and EBSCO Host, including Academic Search Elite and Business Source Elite. In all databases, searches were made in titles, abstracts or topic descriptions for the following key words: robust design methodology, robust design, Taguchi method, quality engineering and robust engineering. The selection of papers found in the databases was done in two steps – first based on title, source and author/s and second on abstract. The selection criteria were derived from the purpose of investigating principles of robust design, i.e. not chiefly focusing on statistical methods useful in robust design. The final number of papers identified in the database searches is given in Table 3.1.

In Table 3.1 a reference is counted only once, even if it was found in multiple databases. In addition to the database searches, book searches have been made in the Swedish national catalogue Libris and the catalogue of Oxford University Library. The initial search was conducted in November 2003, with complementary searches up to November 2006. In addition to these searches, literature was found by following citations in papers and books. From this literature a number of recurring and central aspects of robust design were identified.

Table 3.1 Summary of the database searches; bold figures are the number of selected papers per search item and database; figures in parentheses are the number of hits before any selection.

| | Database | | | |
Search item	Dissertation Abstracts	Compendex	Science Citation Index	EBSCO
Robust design methodology	**2** (16)	**2** (37)	**2** (28)	**3** (7)
Robust design	**3** (178)	**16** (1156)	**26** (797)	**17** (151)
Taguchi method	**2** (53)	**9** (684)	**8** (563)	**8** (82)
Quality engineering	**0** (39)	**7** (385)	**5** (162)	**5** (96)
Robust engineering	**0** (4)	**0** (27)	**0** (7)	**0** (2)

3.3 Results and Analysis

3.3.1 Terminology

The concepts used to describe efforts to reduce variation in product characteristics are many; some examples are *Taguchi methods, quality engineering, parameter design, robust design, robust engineering* and *robust design methodology*. As seen in Table 3.1, *robust design* is the most common term in databases containing papers and dissertations, followed by *Taguchi methods*. Moreover, textbooks on Taguchi methods are quite common. However, they are not included in the table as they were found in book searches.

The concept of *quality engineering* appears in publications such as Taguchi and Wu (1979), Taguchi (1986, 1993) and Saitoh et al. (2003a, 2003b). Taguchi (1993) writes that 'quality engineering is not intended to reduce the sources of variation in products directly. Instead, one needs to make the systems of products or production processes less sensitive to sources of uncontrollable noise, or outside influences, through parameter design (off-line quality control) methods.' In the literature, *quality engineering* encompasses the concepts of system design, parameter design and tolerance design. These concepts are often labelled *Taguchi methods* or *Taguchi techniques*, see Roy (1990), Benton (1991), Ross (1996), and Wu and Wu (2000).

In addition to the concepts dealt with so far there are a number of concepts that are more general than those that indicate a connection to the work of Taguchi, e.g. *robustness, robust design* and *robust design methodology*. These are used in, for example, Goh (1993), Matthiasen (1997), Andersson (1997a), Thornton et al. (2000), Box (1999), Gremyr et al. (2003) and Thornton (2004). These concepts are referred to as general because they do not, in the same way as the concepts related to Taguchi, prescribe specific methods for reducing variation. Rather, they are used when robustness is seen as an engineering problem that can be solved in a number of ways. A *robust design* is of course also used as a description of a characteristic of a product. This characteristic of being robust simply has to do with insensitivity to variation; for discussions and elaborations see Hoehn (1995), Ford (1996) and Andersson (1997a).

3.3.2 View of Variation

Although many authors – for example Tribus and Szonyi (1989), Pignatiello and Ramberg (1991) and the contributors in Nair (1992) – are critical of the statistical methods advocated by Taguchi, they give him credit for his work on communicating the importance of reducing variation to industry. As Lorenzen writes in Nair (1992) '... there is certainly no doubt that Taguchi has popularized the idea of robustness within the engineering community, and this is a big contribution.' This points to a common base in the literature on robust design, the view that variation is a fundamental problem, as stated in Goh (2002): 'Regardless of the physical criteria used as proxy to quality and the way in which specifications are arrived at, all quality problems during generation of goods and services arise fundamentally from only one source: variation.' Similarly, Box and Bisgaard (1988) write 'the enemy of mass production is variability. Success in reducing it will invariably simplify processes, reduce scrap, and lower costs.'

3.3.2.1 The Quadratic Loss Function

Taguchi and Wu (1979) define quality as 'the losses a product imparts to the society from the time the product is shipped'. These losses fall under two categories: loss caused by functional variation and loss caused by harmful side effects. Kackar (1989) questions the definition of quality given above for neglecting losses before a product is shipped, e.g. scrap in manufacturing, and suggests an extension of the definition to include manufacturing loss as well.

A central illustration of the concept of loss used in, for example, Taguchi and Wu (1979), Roy (1990) and Ross (1996) is the quadratic loss function, an illustration that Benton (1991) refers to as 'the heart of the Taguchi philosophy'. A criticism of loss functions, see Matthiasen (1997) and Box et al. (1988), is that it is a good illustration but not actually useful owing to the difficulties of characterizing and balancing economic loss. However, Bergman and Klefsjö (2003) refer to a practical case at General Motors where the quality loss function was estimated through customer ranking of unsatisfactory and satisfactory levels of a certain parameter.

3.3.2.2 Noise Factors

The forces that cause deviation from target and thus induce loss are often labelled noise, noise factors or simply sources of variation, see for example Taguchi (1986), Park (1996), Ross (1996) and Wu and Wu (2000). Many authors divide noise factors into different categories as a means to clarify the kinds of noise that can induce loss, see for example Taguchi and Wu (1979), Taguchi (1986), Clausing (1994) and Phadke (1989). There are most often three categories, the names used for these categories are different but the content of them is very similar. Examples of categories are external noise, internal noise and unit-to-unit noise as proposed in Taguchi (1986) or, as suggested by Clausing (1994), variations in conditions of use, production variations and deterioration (variation with time and use).

Besides inducing loss, noise factors are difficult, expensive or even impossible to control. As stated in, for example, Gunter (1987), Phadke (1989) and Park (1996), noise factors cannot be easily or cost effectively controlled, which makes insensitivity to noise factors rather than the elimination of them the preferable way to achieve a robust design. The final aim, i.e. robustness, is defined in Taguchi et al. (2000) as 'the state where the technology, product, or process performance is minimally sensitive to factors causing variability (either in manufacturing or user's environment) and aging at the lowest manufacturing cost.' In other words, the aim is not to eliminate noise but to create insensitivity to it. This can be done either by identifying levels of control factors that result in a more robust product; or in some cases even by redesign of the product.

3.3.3 Procedures

3.3.3.1 A Specific Procedure or an Overall Approach

Looking briefly at parts of the literature on robust design it is tempting to believe that robust design is synonymous with the three-step procedure proposed by Taguchi (1986) and that it implies the use of designed experiments. This view is illustrated by the following quote from Tsui (1992): 'Robust design is an efficient and systematic methodology that applies statistical experimental design for improving product and manufacturing process design. . . . The robust design method was originally developed by a Japanese quality consultant, Genichi Taguchi.' Shoemaker and Tsui express another view in Nair (1992) 'It should be emphasized that robust design is a problem in product design and manufacturing-process design and that it does not imply any specific solution methods.' The intriguing fact that both these illustrative quotes were taken from the same author in the same year might indicate a certain level of inattention to the concept of robust design.

Mörup (1993), Thornton et al. (2000) and Tennant (2002) consider robust design to be an overall approach and Shoemaker and Tsui in Nair (1992) consider robust design as 'a problem' in engineering. A difference between authors who consider robust design to be a specific procedure and those who view it as an overall approach concerns the concept or system design phase. Anderson (1996a) says that the development of robust designs may be enhanced by appropriate choices of concepts, which, in Taguchi's terminology, is a part of system design. Anderson (1996a) further argues that a wise choice of concept is the only possible option for achieving robust design in situations in which physical experimentation or computer simulations are impossible. Examples of methods proposed by Andersson (1996a, 1996b, 1997b) are an adapted Failure Mode and Effect Analysis, which takes into account the influence of noise factors, the use of the error transmission formula to evaluate different design solutions and the use of design rules and principles that contribute to robust design. Another author that deals with robustness in system design is Goh (1993). His proposal is to simulate environmental variables in the design stage to expose possible weaknesses in a product or process. Taguchi (1993) on the other hand argues that robust design is not dependent to any great extent on work in the system design phase.

The view of robust design as a specific procedure is supported in Lin et al. (1990) and Chowdhury (2002). Here robust design is seen as identical to the application of a designed experiment analysed by signal-to-noise ratios as a means of finding a robust solution. In Chowdhury (2002) robust design is applied as a subordinate part of the overall framework Design for Six Sigma (DFSS). In this case robust design is seen as a synonym of the Taguchi (1986) three-step procedure with a focus on the use of design of experiments in parameter design.

3.3.3.2 Taguchi's Three-Step Procedure

The design process described in Taguchi and Wu (1979) and Taguchi (1985a, 1986) is strongly associated with the concept of quality engineering, see e.g. Ramberg et al. (1991) and Goh (1993). According to the description in Taguchi (1986), system design is the stage in which different concepts and choices of technology are considered at different levels, e.g. the system level and component level. The aim in parameter design is to decide on appropriate levels of individual system parameters. What is an appropriate level for a parameter or an appropriate combination of parameter levels is determined by what reduces the effect of noise on the output characteristic. Finally, in the last step, tolerances are set in a way that further minimizes the effect of noise, e.g. narrower tolerances for noise factors that have the greatest influence. Taguchi (1986) emphasizes, however, that this is not the most efficient way to reduce variation caused by noise and that it should be seen as a last resort after parameter design.

Work to achieve robustness is divided in Taguchi and Wu (1979) into off-line and on-line quality control efforts, where the former is applied during the design of a product and the

Table 3.2 Off-line and on-line quality control as countermeasures against noise (Taguchi, 1986).

Department countermeasures			Type of noise		
			External	Internal	Unit-to-unit
Off-line quality control	Research and development	(1) System design	•	•	•
		(2) Parameter design	•	•	•
		(3) Tolerance design	0	•	•
	Production engineering	(1) System design	X	X	•
		(2) Parameter design	X	X	•
		(3) Tolerance design	X	X	•
On-line quality control	Production	(1) System design	X	X	•
		(2) Parameter design	X	X	•
		(3) Tolerance design	X	X	•
	Customer relations	After-sales service	X	X	X

•, Possible; 0, possible, but should be a last resort; X, impossible.

latter during production. The different approaches vary in their ability to create robustness against different categories of noise factors. The focus in this chapter is what Taguchi refers to as off-line quality control effort. In many references, for example Taguchi (1985b, 1986), Kackar (1985) and Phadke (1989), on-line and off-line quality control are related to categories of noise factors in tables similar to Table 3.2.

The major means of achieving a robust design according to Taguchi (1986) is through parameter design; thus it is less important to think about robustness in system design. Lin et al. (1990) say that Taguchi methods are useful from the beginning of research and development to the end of the production line, but system design is left out in their list of activities. Another example of the focus on parameter design is in Chan and Xiao (1995), who view robust design and parameter design as synonymous. Benton (1991) summarizes this by saying 'System design . . . In this phase Taguchi methods offer little help.' The emphasis on the latter phases of product design is not uncontroversial, as can be seen in Ford (1996): 'the traditional approaches to robust design are limited to optimization of design parameter values and neglect opportunities at the concept definition stage.' Anderson (1996a) argues that Taguchi overemphasizes parameter design as the only option for achieving robust designs. Such authors as Dabade and Ray (1996) and Wilkins (2000) also seem to have interpreted his work in this way.

3.3.4 Objective

While there is a common aim of reducing unwanted variation, there is controversy in the more long-term aims of these efforts. Parr (1988) describes this as two approaches toward robust design: one solution oriented and another understanding oriented. Lin et al. (1990) illustrate the difference between these approaches: 'The difference between statistical methods and Taguchi methods is that statistical methods tell you what has happened'.

The notion of a solution-oriented and an understanding-oriented approach to robust design is described in some of the literature on robust design as an issue of engineering versus science. A statement that clearly illustrates the solution-oriented view is found in Nair (1992), where Shin Taguchi states: 'The goal in parameter design is not to characterize the system but to achieve robust function. Pure science strives to discover the causal relationships and to understand the mechanics of how things happen. Engineering, however, strives to achieve the results needed to satisfy the customer. Moreover, cost and time are very important issues for engineers. Science is to explain nature while engineering is to utilize nature.' Taguchi (1993) and Taguchi et al. (2000) have similar views.

George Box responds to Shin Taguchi in the panel discussion in Nair (1992): '... the ultimate goal must surely be to better understand the engineering system. ... I profoundly disagree with the implications of Shin Taguchi's claim that the engineer does not need to "discover the causal relationship and to understand the mechanics of how things happen".' Other authors that emphasize the understanding of the system under study are Box et al. (1988), Tsui (1996), Scibilia et al. (2001) and Smith and Clarkson (2005). The understanding versus solution oriented approach to robust design is also reflected in the use of experimental designs. Pignatiello and Ramberg (1991) and Goh (1993) criticize Taguchi for reducing the use of experimental design to a set of 'cookbook' procedures that give no explanations of 'why?' and 'how?'.

3.3.5 Methods and Methodologies

Numerous methods have been developed to support the design of robust products. The majority focus on design improvement and only a limited number have been proposed as an aid to support robust design in what Taguchi and Wu (1979) refer to as system design. However, Thornton (2004) proposes a methodology called Variation Risk Management (VRM) that is intended to be incorporated in a product development process. According to Thornton (2004), VRM can serve as an overall framework for reducing variation from system design to production.

Johansson et al. (2006) describe a method for variation reduction that is applicable in the concept selection phase as well as in improving existing designs. They call the method Variation Mode and Effects Analysis (VMEA) and maintain that it is useful for identifying noise factors that influence a design. Further, Anderson (1996a) suggests a semi-analytic approach based on the error transmission formula (see e.g. Morrison, 1957) that may serve as an aid in comparisons of the sensitivity to noise factors in different designs. Matthiasen (1997) and Andersson (1997b) suggest the use of design principles as a means of imposing robust design in system design phases. Another method for robust design in the concept stages of development is the robust concept design method developed by Ford (1996). This consists of four steps: defining the robustness problem, deriving guiding principles, making a new concept synthesis and evaluating alternative concepts.

Interest in robust design, and particularly robust design experimentation, increased in the USA and in Europe in the 1980s, partly owing to the publishing of Taguchi (1986). Taguchi (1986) advocates the use of orthogonal arrays where both control factors and noise factors are varied. These experiments are analysed by the use of signal-to-noise ratios to identify robust designs. The rationale behind the signal-to-noise ratios is that they are linked to the quadratic loss function. Maghsoodloo (1990) explores the exact relation between some of the most commonly used signal-to-noise ratios and the quadratic loss function.

The essence of Taguchi's ideas on the use of design of experiments to improve designs was explained by Kackar (1985) and Hunter (1985). A large number of papers have been published on the use of design of experiments to identify robust designs, and a recent review by Robinson et al. (2006b) summarizes the central points in the research in this area. However, as Li et al. (1998) point out, there is a need for more work on other methods than design of experiments in the parameter design phase. Examples are the use of smart assemblies described in Downey et al. (2003), standardization of parts as a means of withstanding production variations outlined in Little and Singh (1996), and the integration of a product family concept and robust design suggested in Sopadang (2001).

It is impossible or too expensive in some cases to conduct physical experiments. In such situations it can instead be possible to build simulation models that take into account both control factors and noise factors, see Welch et al. (1990), Benjamin et al. (1995), Ramberg et al. (1991), Barone and Lanzotti (2002), Jeang and Chang (2002) and Creighton and Nahavandi (2003). The influence of noise factors on the system can be investigated in appropriately planned computer experiments.

When variation cannot be brought to a satisfactory level by wisely selected control factor levels, Taguchi and Wu (1979) argue that it should be minimized by controlling the noise factors. Taguchi (1986) writes that choosing the noise factors to control should be based on the associated cost and their contribution to the total amount of variation. Nigam and Turner (1995) present a review of statistically based methods for studying how variation caused by different noise factors propagates to the total variation. Bisgaard and Ankenman (1995) argue that the Taguchi two-step strategy of first seeking appropriate control factor settings and then, when needed, controlling certain noise factors may lead to non-optimal solutions. Chan and Xiao (1995) and Li and Wu (1999) integrate these two steps for reducing variation.

3.3.6 Experimental Approach

The core of Taguchi's parameter design is based on experimental methods. Since the beginning of the 1980s there has been ongoing research to suggest alternatives and improvements of the methods he suggested. Useful references to obtain a picture of the current status of robust parameter design (RPD) are Wu and Hamada (2000), Myers and Montgomery (2002), Robinson et al. (2003) and Myers et al. (2004).

3.3.6.1 Suitable Experimental Designs

The basic idea of RPD experiments is to vary control factors and noise factors in the same experiment and seek possible control-noise factor interactions. The combined array approach suggested by Welch et al. (1990) and Shoemaker et al. (1991) includes noise factors and control factors in the same design matrix, which often results in cost-efficient experiments. Moreover, by use of mixed resolution designs suggested by Lucas (1989), experiments can be designed with high resolution for control–noise factor interactions and control–control factor interactions while having lower resolution for noise–noise factor interactions. Lucas (1994) proposed use of response surface designs with mixed resolution and emphasized their superiority over the experimental designs advocated by Taguchi. Borkowski and Lucas (1997) continued the work in Lucas (1994) and developed a catalogue of mixed resolution response surface designs.

Another resource efficient experimental arrangement is split-plot experiments where the experimental order is not completely randomized and all factor levels are not reset between each experiment, see Box and Jones (1992), Letsinger et al. (1996) and Bisgaard (2000). Another positive characteristic of these experiments is that they allow a precise estimation of control–noise factor interactions. Bisgaard (2000) showed how the mixed resolution concept can also be applied to split-plot experiments. The use and usefulness of split-plot designs in RPD have been discussed by Kowalski (2002), Loeppky and Sitter (2002), Bingham and Sitter (2001, 2003) and McLeod and Brewster (2006).

Borror et al. (2002) evaluated design matrices for RPD with respect to two different variance criteria. Other recent references suggesting design evaluation criteria are Bingham and Li (2002) and Loeppky et al. (2006).

3.3.6.2 Different Approaches to Analysis

Robinson et al. (2003) and Myers et al. (2004) discuss the two response surface methodology (RSM) approaches that evolved during the 1990s to analyse robust design problems: the single and the dual response approaches. In the dual model approach, originally proposed by Vining and Myers (1990), separate models are fitted for the mean and the variance. This type of analysis requires replicated experiments or the inner–outer arrays type experiments often advocated by Taguchi (1986). In the single response approach originally proposed by Welch et al. (1990), control factors and noise factors coexist in the same model. Myers et al. (1992) pointed out that two response surfaces, one for the mean and one for the variance, can be determined theoretically from the single response model. The two models are further considered jointly to identify the best choice of control factor settings. Steinberg and Bursztyn (1998) compared the single and dual response approaches and showed that the single response approach is favourable if noise factors have been controlled at fixed levels in experiments.

Lee and Nelder (2003) proposed a Generalized Linear Model (GLM) approach when common assumptions made in RSM are not justified. Engel and Huele (1996) proposed a GLM for situations where the residual variance cannot be assumed to be constant. Myers et al. (2005) suggested a GLM approach for the analysis of RPD experiments with non-normal responses. Robinson et al. (2006a) extended the analysis to experiments with random noise variables and non-normal responses, in this case Generalized Linear Mixed Models should be used as GLMs require independent responses.

Brenneman and Myers (2003) investigated the situation when categorical noise factors, as for example different performance of production equipment or different suppliers, are varied in RPD experiments. Under the assumption that the proportions of each category of the noise factors are known, an analysis procedure was proposed. Moreover, Brenneman and Myers (2003) implied that process variance can be reduced by utilizing not only possible control–noise factor interactions but also by adjusting the proportions of the categories of the noise factors. Robinson et al. (2006a) continued the work of Brenneman and Myers (2003) and proposed an analysis strategy taking into account the possibility to make adjustments of the category proportions and the cost or suitability associated with different control factor settings.

The experimental designs and analyses procedures dealt with in this section are all based on explicit modelling of the response and aimed at increasing the understanding of the problem under study. In other words, current research on robust design is in line with the understanding-oriented approach rather than the solution-oriented approach (see Section 3.3.4).

3.4 Discussion

It is clear from the literature reviewed in this chapter that there are many, almost synonymous concepts that are used to describe efforts to achieve robust designs. However, there is agreement on some fundamental principles among the users of different concepts. This common ground can be found in the view of variation and the emphasis on insensitivity to, rather than elimination of, noise factors. It is shown in Gremyr et al. (2003) that practitioners also recognize variation as a problematic issue. Thus it is important to further investigate why the research conducted on robust design does not seem to meet this industrial need.

A survey of Swedish manufacturing companies presented in Arvidsson et al. (2003) and Gremyr et al. (2003) shows that as few as 28% of the companies in that study recognize the concept of robust design despite the fact that the area has received considerable interest in journals and textbooks. It has also been observed that industrial users of robust design emphasize the need for support in applying RDM. Taking these studies into consideration, the inattention to the framework of RDM in favour of research on applicable statistical methods might have contributed to the low level of industrial use.

The study of the literature treated in this chapter brings out the importance of reflecting upon a number of specific areas that have yet received little or no attention. One is the implementation of RDM. How should a company that wants to start working with RDM proceed? Two publications that are related to this topic are by Saitoh et al. (2003a, 2003b), which describe how quality engineering has been implemented in Fuji Zerox in Japan. Concerning the ideas in robust design that have their origin in Japan, one thing that must be considered when taking it to a western company is, as discussed in Goh (1993), are cultural differences. Another aspect of implementation is the organization of robust design work; should it be run in individual projects or as an integrated part of a company's development process? Finally, it seems that efforts are needed in developing methods that support awareness of variation in early design phases, e.g. in concepts generation.

3.5 Conclusions

3.5.1 Synthesis

Despite all the debates on the practical use of the quality loss function, the most fundamental common denominator in the literature on robustness and robust design is that unwanted variation imparts loss. This loss is induced by noise factors that are often hard or impossible to control in an affordable way. These factors can arise in the conditions of use or in production or be caused by deterioration. Due to the nature of noise factors the aim of robust design is to create insensitivity to them rather than try to eliminate or control them.

A more controversial area is the approaches for achieving robust design. For some authors robust design is a general problem, whereas others see it as synonymous to Taguchi's three-step procedure. Further, there is disagreement on the value of robust design efforts in the development and selection of concepts, where some argue that it is crucial to apply robust design as early as possible whereas others feel that it is applicable mainly in the realization of a chosen concept.

There is no conflict over design of experiments being useful in robust design; the way in which these experiments should be conducted and analysed, however, is the greatest point of conflict in the literature on robust design. It concerns various factors, such as resource efficiency, view on interactions, signal-to-noise ratios and one-shot versus sequential experimentation. The latter aspect is also closely connected to another controversy: whether the goal of robust design is to arrive at one robust solution or whether it is also a matter of understanding the underlying causes of variation. Finally, despite the many controversies over the statistically based methods advocated by Taguchi, there is broad agreement on the value of Taguchi's contribution in emphasizing variation reduction and creating industrial interest in it.

3.5.2 A Definition of Robust Design Methodology

This examination of the literature review raises a fundamental question: what is robust design? It is apparent from the sections above that this question has many different answers. Below we will give our answer.

First, a cornerstone of robust design is an awareness of variation. Talking about products, there is a view in which the product is affected not only by controllable factors but also by factors that are uncontrollable or hard to control. The latter factors are referred to as noise factors, which cause a characteristic to deviate from its desired and/or specified level. Until now most research on variation reduction has focused on method development rather than on increasing and emphasizing the awareness of variation. This focus might be one factor underlying the low level of industrial applications of robust design, as reported in Gremyr et al. (2003). Further, there would probably be a better understanding of the methods that have been developed if the underlying views on variation were better articulated.

Second, the nature of noise factors as being impossible or expensive to control often makes robustness by means of control or elimination of these factors an unappealing approach. The goal in robust design is rather to create insensitivity to noise factors.

Third, we argue that robust design does not in itself prescribe the use of certain methods applied in specified steps. This should not be interpreted, however, as though it only has to do with creating a robust product in any possible way by making it stronger and heavier, but through insensitivity to, and awareness of, noise factors. To stress this we would like to use the concept of robust design methodology (RDM). Methodology is defined in the Oxford English Dictionary as 'a method or body of methods used in a particular field of study or activity'. In our view the field of RDM is based on a number of underlying models: a product as being affected by controllable factors as well as noise factors, the quality loss function showing that unwanted variation induces loss, and the view that it is important to create insensitivity to all categories of noise factors and to apply such efforts from system design to production. With this base, a number of different methods, not necessarily design of experiments although useful, can be applied to arrive at a robust design. An advantage of using RDM rather than robust design to describe efforts towards robustness is that it distinguishes the methodology from a characteristic. In other words, it clarifies that a product can be robust with or without the application of RDM.

Fourth, RDM is useful from concept generation to the production of a product. Moreover, it is necessary to apply RDM throughout a development process in order to exploit all possibilities for robustness covering all categories of noise factors. The four key areas of RDM can be summarized in the following definition:

Robust Design Methodology means systematic efforts to achieve insensitivity to noise factors. These efforts are founded on an awareness of variation and can be applied in all stages of product design.

References

Andersson, P. A semi-analytic approach to robust design in the conceptual design phase. *Research in Engineering Design*, **8**: 229–239, 1996a.

Andersson, P. A practical method for noise identification and failure mode importance ranking in robust design engineering. In *III International Congress of Project Engineering*, Barcelona, 1996b.

Andersson, P. Robustness of technical systems in relation to quality, reliability and associated concepts. *Journal of Engineering Design*, **8**: 277–288, 1997a.

Andersson, P. On robust design in the conceptual design phase – a qualitative approach. *Journal of Engineering Design*, **8**: 75–90, 1997b.

Arvidsson, M., Gremyr, I. and Johansson, P. Use and knowledge of robust design – a survey of Swedish industry. *Journal of Engineering Design*, **14**: 1–15, 2003.

Barone, S. and Lanzotti, A. Quality engineering approach to improve comfort of a new vehicle in virtual environment. In *Proceedings of the American Statistical Association Conference*, Alexandria, 2002.

Benjamin, P. C., Erraguntla, M. and Mayer, R. J. Using simulation for robust system design. *Simulation*, **65**: 116–128, 1995.

Benton, W. C. Statistical process control and the Taguchi method: a comparative evaluation. *International Journal of Production Research*, **29**: 1761–1770, 1991.

Bergman, B. and Klefsjö, B. *Quality from Customer Needs to Customer Satisfaction*. Studentlitteratur, Lund, 2003.

Bingham, D. and Li, W. A class of optimal robust parameter designs. *Journal of Quality Technology*, **34**: 244–259, 2002.

Bingham, D. and Sitter, R. R. Design issues in fractional factorial split-plot experiments. *Journal of Quality Technology*, **33**: 2–15, 2001.

Bingham, D. and Sitter, R. R. Fractional factorial split-plot designs for robust parameter experiments. *Technometrics*, **45**: 80–89, 2003.

Bisgaard, S. The design and analysis of $2^{k-p} \times 2^{q-r}$ split plot experiments. *Journal of Quality Technology*, **32**: 39–56, 2000.

Bisgaard, S. and Ankenman, B. Analytic parameter design. *Quality Engineering*, **8**: 75–91, 1995.

Borkowski, J. J. and Lucas, J. M. Designs of mixed resolution for process robustness studies. *Technometrics*, **39**: 63–70, 1997.

Borror, C. M., Montgomery, D. C. and Myers, R. H. Evaluation of statistical designs for experiments involving noise variables. *Journal of Quality Technology*, **34**: 54–70, 2002.

Box, G. and Bisgaard, S. Statistical tools for improving designs. *Mechanical Engineering*, **110**: 32–40, 1988.

Box, G. E. P. Statistics as a catalyst to learning by scientific method part II – a discussion. *Journal of Quality Technology*, **31**: 16–29, 1999.

Box, G. E. P. and Jones, S. Split-plot designs for robust experimentation. *Journal of Applied Statistics*, **19**(1): 3–25, 1992.

Box, G. E. P., Bisgaard, S. and Fung, C. An explanation and critique of Taguchi's contributions to quality engineering. *Quality and Reliability Engineering International*, **4**: 123–131, 1988.

Brenneman, W. A. and Myers, R. H. Robust parameter design with categorical noise factors. *Journal of Quality Technology*, **35**: 335–341, 2003.

Chan, L. K. and Xiao, P. H. Combined robust design. *Quality Engineering*, **8**: 47–56, 1995.

Chowdhury, S. *Design For Six Sigma – The Revolutionary Process for Achieving Extraordinary Profits*. Dearborn Trade Publishing, Chicago, 2002.

Clausing, D. *Total Quality Development – A Step-By-Step Guide to World-Class Concurrent Engineering*. ASME Press, New York, 1994.

Creighton, D. and Nahavandi, S. Application of discrete event simulation for robust system design of a melt facility. *Robotics and Computer Integrated Manufacturing*, **19**: 469–477, 2003.

Dabade, B. M. and Ray, P. K. Quality engineering for continuous performance improvement in products and processes: A review and reflections. *Quality and Reliability Engineering International*, **12**: 173–189, 1996.

Downey, K., Parkinson, A. and Chase, K. An introduction to smart assemblies for robust design. *Research in Engineering Design*, **14**: 236–246, 2003.

Engel, J. and Huele, A. F. A generalized linear modeling approach to robust design. *Technometrics*, **38**: 365–373, 1996.

Ford, R. B. *Process for the Conceptual Design of Robust Mechanical Systems – Going Beyond Parameter Design to Achieve World-Class Quality.* Stanford University, Palo Alto, 1996.

Goh, T. N. Taguchi methods: Some technical, cultural and pedagogical perspectives. *Quality and Reliability Engineering International,* **9**: 185–202, 1993.

Goh, T. N. The role of statistical design of experiments in six sigma: Perspectives of a practitioner. *Quality Engineering,* **4**: 659–671, 2002.

Gremyr, I., Arvidsson, M. and Johansson, P. Robust design methodology: Status in the Swedish manufacturing industry. *Quality and Reliability Engineering International,* **19**: 285–293, 2003.

Grize, Y. L. A review of robust design approaches. *Journal of Chemometrics,* **9**: 239–262, 1995.

Gunter, B. A perspective on the Taguchi method. *Quality Progress,* **June**: 44–52, 1987.

Hoehn, W. K. Robust designs through design to six sigma manufacturability. In *Proceedings of the IEEE International Engineering Management Conference,* Singapore, pp. 241–246, 1995.

Hunter, J. S. Statistical design applied to product design. *Journal of Quality Technology,* **17**: 210–221, 1985.

Jeang, A. and Chang, C. Combined robust parameter and tolerance design using orthogonal arrays. *International Journal of Advanced Manufacturing Technology,* **19**: 442–447, 2002.

Johansson, P., Chakhunashvili, A., Barone, S. and Bergman, B. Variation mode and effect analysis: a practical tool for quality improvement. *Quality and Reliability Engineering International,* **22**: 865–876, 2006.

Kackar, R. N. Off-line quality control, parameter design, and the Taguchi method (with discussion). *Journal of Quality Technology,* **17**: 176–188, 1985.

Kackar, R. N. *Taguchi's Quality Philosophy: Analysis and Commentary. Quality Control, Robust Design, and the Taguchi Method.* Wadsworth & Brooks/Cole Advanced Books & Software, Pacific Grove, CA, 1989.

Kowalski, S. M. 24 run split-plot experiments for robust parameter design. *Journal of Quality Technology,* **34**: 399–410, 2002.

Lee, Y. and Nelder, J. A. Robust design via generalized linear models. *Journal of Quality Technology,* **35**: 2–12, 2003.

León, R. V., Shoemaker, A. C. and Kackar, R. N. Performance measures independent of adjustment. *Technometrics,* **29**: 253–285, 1987.

Letsinger, D. L., Myers, R. H. and Lentner, M. Response surface methods for bi-randomization structures. *Journal of Quality Technology,* **28**: 381–397, 1996.

Li, C. C., Sheu, T. S. and Wang, Y. R. Some thoughts on the evolution of quality engineering. *Industrial Management and Data Systems,* **98**: 153–157, 1998.

Li, W. and Wu, C. F. J. An integrated method of parameter design and tolerance design. *Quality Engineering,* **11**: 417–425, 1999.

Lin, P., Sullivan, L. and Taguchi, G. Using Taguchi methods in quality engineering. *Quality Progress,* **23**: 55–59, 1990.

Little, T. A. and Singh, G. 10 keys to achieving robust product and process design. In *Proceedings of the Annual Quality Congress,* Chicago, ASQC, pp. 370–376, 1996.

Loeppky, J. L. and Sitter, R. R. Analyzing un-replicated blocked or split-plot fractional factorial designs. *Journal of Quality Technology,* **34**: 229–243, 2002.

Loeppky, J. L., Bingham, D. and Sitter, R. R. Constructing non-regular robust parameter designs. *Journal of Statistical Planning and Inference,* **136**: 3710–3729, 2006.

Lucas, J. M. Achieving a robust process using response surface methodology. In *Proceedings of the American Statistical Association Conference,* Washington, DC, pp. 579–593, 1989.

Lucas, J. M. How to achieve a robust process using response surface methodology. *Journal of Quality Technology,* **26**: 248–260, 1994.

Maghsoodloo, S. The exact relation of Taguchi's signal-to-noise ratio to his quality loss function. *Journal of Quality Technology,* **22**: 57–67, 1990.

Matthiasen, B. *Design for robustness and reliability – Improving the quality consciousness in engineering design.* PhD thesis, Technical University of Denmark, Lyngby, 1997.

McLeod, R. G. and Brewster, J. F. Blocked fractional factorial split-plot experiments for robust parameter design. *Journal of Quality Technology,* **38**: 267–279, 2006.

Morrison, J. S. The study of variability in engineering design. *Applied Statistics,* **6**: 133–138, 1957.

Mörup, M. *Design for Quality.* Institute for Engineering Design, Technical University of Denmark, Lyngby, 1993.

Myers, R. H. and Montgomery, D. C. *Response Surface Methodology – Process and Product Optimization Using Designed Experiments.* John Wiley & Sons Ltd, New York, 2002.

Myers, R. H., Khuri, A. I. and Vining, G. G. Response surface alternatives to the Taguchi robust parameter design approach. *The American Statistician*, **46**: 131–139, 1992.

Myers, R. H., Montgomery, D. C., Vining, G. G., Borror, C. M. and Kowalski, S. M. Response surface methodology: A reterospective and litterature survey. *Journal of Quality Technology*, **36**: 53–77, 2004.

Myers, R. M., Brenneman, W. A. and Myers, R. H. A dual-response approach to robust partameter design for a generalized linear model. *Journal of Quality Technology*, **37**: 130–138, 2005.

Nair, V. N. Taguchi's parameter design: A panel discussion. *Technometrics*, **34**: 127–161, 1992.

Nigam, S. D. and Turner, J. U. Review of statistical approaches to tolerance analysis. *Computer-Aided Design*, **27**: 6–15, 1995.

Park, G., Lee, T., Lee, K. and Hwang, K. Robust design: An overview. *AIAA Journal*, **44**: 181–191, 2006.

Park, S. H. *Robust Design and Analysis for Quality Engineering*. Chapman & Hall, London, 1996.

Parr, W. C. Discussion to signal-to-noise ratios, performance criteria, and transformations. *Technometrics*, **30**: 22–23, 1988.

Phadke, M. S. *Quality Engineering using Robust Design*. Prentice Hall, Englewood Cliffs, NJ, 1989.

Pignatiello, J. J. and Ramberg, J. S. Top ten triumphs and tragedies of Genichi Taguchi. *Quality Engineering*, **4**: 211–225, 1991.

Ramberg, J. S., Sanchez, S. M., Sanchez, P. J. and Hollick, L. J. Designing simulation experiments: Taguchi methods and response surface metamodels. In *Proceedings of the Winter Simulation Conference*, Phoenix, USA, pp. 167–176, 1991.

Robinson, T. J., Borror, C. M. and Myers, R. H. Robust parameter design: A review. *Reliability Engineering International*, **20**: 81–101, 2003.

Robinson, T. J., Brenneman, W. A. and Myers, R. M. Process optimization via robust parameter design when categorical noise factors are present. *Quality and Reliability Engineering International*, **22**: 307–320, 2006a.

Robinson, T. J., Wulff, S. S., Montgomery, D. C. and Khuri, A. I. Robust parameter design using generalized linear mixed models. *Journal of Quality Technology*, **38**: 65–75, 2006b.

Ross, P. J. *Taguchi Techniques for Quality Engineering*. McGraw-Hill, New York, 1996.

Roy, R. *A Primer on the Taguchi Method*. Society of Manufacturing Engineers, Dearborn, MI, 1990.

Saitoh, K., Yoshizawa, M., Tatebayashi, K. and Doi, M. A study about how to implement quality engineering in research and development (Part 1). *Journal of Quality Engineering Society*, **11**: 100–107, 2003a.

Saitoh, K., Yoshizawa, M., Tatebayashi, K. and Doi, M. A study about how to implement quality engineering in research and development (Part 2). *Journal of Quality Engineering Society*, **11**: 64–69, 2003b.

Scibilia, B., Kobi, A., Chassagnon, R. and Barreau, A. Robust design: A simple alternative to Taguchi's parameter design approach. *Quality Engineering*, **13**: 541–548, 2001.

Shainin, D. and Shainin, P. Better than Taguchi orthogonal arrays. *Quality and Reliability Engineering International*, **4**: 143–149, 1988.

Shoemaker, A. C., Tsui, K. L. and Wu, J. Economical experimentation methods for robust design. *Technometrics*, **33**(4): 415–427, 1991.

Smith, J. and Clarkson, P. J. A method for assessing the robustness of mechanical designs. *Journal of Engineering Design*, **16**: 493–509, 2005.

Sopadang, A. *Synthesis of product family-based robust design: Development and analysis*. PhD thesis, Clemson University, Clemson, SC, 2001.

Steinberg, D. M. and Bursztyn, D. Noise factors, dispersion effects, and robust design. *Statistica Sinica*, **8**: 67–85, 1998.

Taguchi, G. Quality engineering in Japan. *Bulletin of the Japan Society of Precision Engineering*, **19**: 237–242, 1985a.

Taguchi, G. Quality engineering in Japan. *Communications in Statistics – Theory and Methods*, **14**: 2785–2801, 1985b.

Taguchi, G. *Introduction to Quality Engineering – Designing Quality into Products and Processes*. Asian Productivity Organization, Tokyo, 1986.

Taguchi, G. *Taguchi on Robust Technology Development – Bringing Quality Engineering Upstream*. ASME Press, New York, 1993.

Taguchi, G. and Wu, Y. *Introduction to Off-Line Quality Control*. Central Japan Quality Control Association, Nagoya, Japan, 1979.

Taguchi, G., Chowdhury, S. and Taguchi, S. *Robust Engineering – Learn How to Boost Quality while Reducing Costs and Time to Market*. McGraw-Hill: New York, 2000.

Tennant, G. *Design for Six Sigma: Launching New Products and Services Without Failure*. Gower Publishing Limited, Hampshire, 2002.

Thornton, A. C. *Variation Risk Management: Focusing Quality Improvements in Product Development and Production*. John Wiley & Sons, Ltd, Chichester, 2004.

Thornton, A. C., Donnely, S. and Ertan, B. More than just robust design: Why product development organizations still contend with variation and its impact on quality. *Research in Engineering Design*, **12**: 127–143, 2000.

Tribus, M. and Szonyi, G. An alternative view of the Taguchi approach. *Quality Progress*, **22**: 46–52, 1989.

Tsui, K. A critical look at Taguchi's modelling approach for robust design. *Journal of Applied Statistics*, **23**: 81–95, 1996.

Tsui, K. L. An overview of Taguchi method and newly developed statistical methods for robust design. *IIE Transactions*, **24**: 44–57, 1992.

Vining, G. G. and Myers, R. H. Combining Taguchi and response-surface philosophies – a dual response approach. *Journal of Quality Technology*, **22**: 38–45, 1990.

Welch, W. J., Yu, T. K., Kang, S. M. and Sacks, J. Computer experiments for quality control by parameter design. *Journal of Quality Technology*, **22**: 15–22, 1990.

Wilkins, J. O. Jr. Putting Taguchi methods to work to solve design flaws. *Quality Progress*, **33**: 55–59, 2000.

Wu, C. F. J. and Hamada, M. *Experiments – Planning, Analysis, and Parameter Design Optimization*. John Wiley & Sons, Inc., New York, 2000.

Wu, Y. and Wu, A. *Taguchi Methods for Robust Design*. The American Society of Mechanical Engineers: New York, 2000.

Part Two

Methods

Methods for taking variation and uncertainty into account in the engineering design process include (1) organizational tools for a rational handling of influential sources, and (2) tools for quantitative measures of the uncertainty in strength, life, or malfunction. The chapters in this part start with the qualitative aspect of including variation and continue with methods of quantitative estimates of different degrees of complexity.

Chapter 4, 'Including Noise Factors in Design Failure Mode and Effect Analysis', takes the industrial practise of using the FMEA tool as its starting point. The potential causes of failure identified by the FMEA procedure are here related to variation and classified in accordance with the noise factor categories, originally defined by Taguchi, and further extended by Davis. From a set of existing FMEA analyses, it is concluded that most of the potential causes of failure can be classified according to the chosen scheme, and thereby be subject to more deep investigations with respect to variability.

Chapter 5, 'Robust Product Development using Variation Mode and Effect Analysis' introduces the concept VMEA, which is an extension of the FMEA tool, taking quantitative measures of failure causes into account. The VMEA method is presented on three different levels of complexity, starting with an engineering judgement of the variable sizes by a simple ranking system and concluding with the probabilistic VMEA, which assigns measures to the different sources by means of the well-defined property: statistical variance. This highest level of complexity corresponds to second moment reliability methods in structural mechanics.

The next two chapters, 6 and 7, use second moment reliability methods for fatigue problems, i.e. they are applications of the probabilistic VMEA. The first one, 'Variation Mode and Effect Analysis: An Application to Fatigue Life Prediction', is the result of a cooperation between statisticians and mechanical engineers with the aim of finding proper safety factors for air engine components. The vague knowledge about the influential variables in the fatigue phenomenon makes it necessary to take model uncertainties into account in the reliability measure, and an essential part of the methodology is to include these unknown errors in the statistical framework.

Robust Design Methodology for Reliability: Exploring the Effects of Variation and Uncertainty
edited by B. Bergman, J. de Maré, S. Lorén, T. Svensson
© 2009, John Wiley & Sons, Ltd

The second fatigue application, 'Predictive Safety Index for Variable Amplitude Fatigue Life', takes the second moment method a step further, and applies the safety index concept to fatigue problems. The well-known concept of load–strength interaction is interpreted by means of the cumulative damage theory and the possible model errors are included in such a way that the usual uncertainties appearing in engineering practise are easily handled.

Chapter 8 compares the second moment approach to a method of higher complexity, namely a Monte Carlo simulation method: 'Monte Carlo Simulations Versus Sensitivity analysis'. An example of the application of the Monte Carlo method for a rather complicated fatigue life calculation is re-examined and compared with the much simpler second moment approach. It turns out that in the particular case the differences are negligible and it is emphasized that the higher complexity model demands detailed knowledge about the input variables to be meaningful.

4

Including Noise Factors in Design Failure Mode and Effect Analysis (D-FMEA) – A Case Study at Volvo Car Corporation

Åke Lönnqvist

4.1 Introduction

Reliability engineering, off-line quality control and Robust Design Methodology (RDM) are approaches, with differences but also similarities, all aiming to avoid failures. Since the main objective is the same, the question is how the methodologies can benefit from each other.

Many well-known methods are offered by reliability engineering, one of them being Failure Mode and Effect Analysis (FMEA), probably the best known reliability method in the automotive industry. This method is integrated in many product development systems and known by many engineers. From a method point of view, FMEA is categorized as a 'bottom-up' method, which means that it identifies causes to failures/failure modes and analyses their potential consequences or effects (Britsman et al., 1993; IEC60300-3-1, 2003; ASQC/AIAG, 2001).

Off-line quality control (Taguchi and Wu, 1979) and RDM (Arvidsson and Gremyr, 2005) offer a way of thinking about the emergence of failures, provided an acceptable 'system design' is available, as the result of noise factor influence. The idea is that for failure-free operation, products must be made insensitive to influence from noise factors.

If noise factors can be interpreted as the causes of failures, the reliability method FMEA and RDM handle the same technical issues. The purpose of this study is to evaluate whether

Robust Design Methodology for Reliability: Exploring the Effects of Variation and Uncertainty
edited by B. Bergman, J. de Maré, S. Lorén, T. Svensson
© 2009, John Wiley & Sons, Ltd

a standard[1] FMEA can be used to study noise factor influence by relating identified causes of failures to noise factor categories. If this holds in reality in the automotive industry, how can this fact be utilized in the FMEA to improve the analysis result and support both reliability engineering and RDM?

The idea to connect FMEAs and RDM is not new; Bergman (1992) proposed to use FMEA as a means to identify noise factors as a preparation for robustness Design of Experiment (DOE). A modified FMEA approach to identify sources of noise is outlined by Andersson (1996). In a step-by-step procedure, noise factors are identified with regard to the requirements area and the life phase. The next step is a transition from noise sources to failure modes and failure effects, similar to ordinary FMEAs with 'causes' replaced by 'noise factors'. Another example is provided by the Variation Mode and Effect Analysis (VMEA) method, which basically utilizes the FMEA form but is a top-down approach, from Key Product Characteristics (KPC), sub-KPCs to noise factors, in order to manage sources of variation (Chakhunashvili et al., 2004). The use of reliability methods such as FMEA and Fault Tree Analysis (FTA) in robust engineering is also discussed by Clausing (1994). It is interesting to note that even if Taguchi does not pay very much attention to (FMEA), his main method being DOE, many textbooks on Design for Six Sigma (DFSS), where robustness is an essential part, mention FMEAs as one of the most important methods for creating robust designs (Lönnqvist, 2006).

4.2 Background

Reliability (performance) is defined as 'the ability of an item to perform a required function under given conditions for a given time interval'. A central role in reliability engineering is the concept of 'failure' and the efforts to avoid causes and consequences. This effort is illustrated by the columns in the FMEA form. The approach in this study is to use the *ordinary standard* FMEA form and study a proposal for how the method can handle noise factor influence.

The FMEA form is based on a number of columns that are connected to the analysis procedure and describes issues to be analysed (Figure 4.1). The order of the columns can differ from form to form, but the analysis procedure is similar. There are, for example, columns for 'potential cause', 'failure mode' and 'effect of failure'. Causes are the origin of failure modes, which in their turn result in failure effects/consequences, as shown in Figure 4.2.

No column in the existing FMEA forms explicitly describes noise factors, but the assumption in this study is that noise factors are actually similar to causes; it is just another way of expressing the same phenomenon.

If a product is sensitive to a noise factor, its function can be negatively influenced and the performance can degrade to such a level that it is regarded as being defective, i.e. there is a product 'soft failure', or, if the product function ceases, the result is a 'hard failure' (Davis, 2004). The hypothesis is that noise factors create conditions that are the causes of failure modes, so specific noise factors and causes are closely linked, or equal, to each other.

In Off-Line Quality Control, Taguchi and Wu (1979) classify noise factors into three categories: outer noise (e.g. temperature, humidity, voltage), inner noise or deteriorating noise (e.g. changes in the properties of the variables of a product during usage) and variational noise

---

[1] Standard FMEA is here referred to as the ASQC/AIAG verson.

Volvo Cars (above)

ARTIKEL/PART		FELKARAKTERISTIK/ CHARACTERISTICS OF FAILURE				RATING					ÅTGÄRD–STATUS/ACTION–STATUS					
Nr/No	Funktion/ Function	Feltyp/ Failure mode	Felorsak/ Causes of failure	Feleffekt på komponent/syst Effects of failure on part/syst	Provning Testing	Po	S	Pd	RPN	Rekommenderad åtgärd/ Recommendations	Beslutad åtgärd/ Decisions taken	Po	S	Pd	RPN	Ansvar avd/sign

QS9000 (below)

Item Function	Potential Failure Mode	Potential Effect(s) of Failure	S e v	C l a s s	Potential Cause(s)/ Mechanism(s) of Failure	O c c u r	Current Design Controls	D e t e c	R. P. N.	Recommended Action(s)	Responsibility & Target Completion Date	Action Results				
												Actions Taken	S e v	O c c	D e t	R. P. N.

Figure 4.1 Examples of columns in Design Failure Mode and Effect Analysis (D-FMEA) forms: Volvo Cars (above) and QS9000 (below).

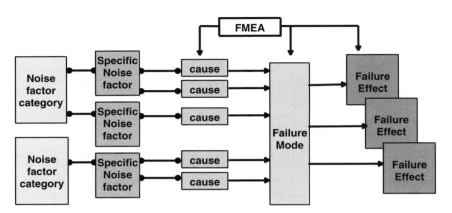

Figure 4.2 A specific noise factor/cause creates conditions that cause a failure mode. Specific noise factors/causes can be grouped into the noise factor categories.

or in-between product noise (production variations). In this study noise factors will not be divided into the three categories above, but into five categories, according to the Ford Motor Company's approach (Davis, 2004). The reason for this is to provide a more distinct and detailed classification when discussing noise factors and potential causes further.

Two of the noise factors influence the possibilities (capability) of the products to withstand the loads (demands) placed upon them. These are the 'internal' or 'capacity' noise factors defined as:

- piece-to-piece variation (production variation)
- changes over time (fatigue, wear out, . . .).

The load, or demand, placed upon a product, depends on the 'external' or 'demand' noise factors, which are:

- customer usage (duty cycles, usage, . . .)
- external environment (temperature, road conditions, . . .)
- systems interaction.

4.3 Method

There are two major kinds of FMEAs, the Design-Failure Mode and Effect Analysis (D-FMEA), which is used to analyse a design, and the Process-Failure Mode and Effect Analysis (P-FMEA), which is used to analyse a process (Britsman et al., 1993). This study covers only D-FMEAs. If the D-FMEA method is applied to a high-level system design, it is here denoted System-FMEA (S-FMEA) instead of D-FMEA. The term D-FMEA is then used for D-FMEAs on the component or subsystem level.

Approximately 30 D-FMEAs/S-FMEAs that have already been performed were selected for the study. All the analyses selected were made during the same period (second quarter of 2006) to diminish any effect of changes in analysis procedures. The intention was to cover all design areas, but it turned out that the selected analysis of the electrical design area had a somewhat different purpose and approach, being high-level S-FMEAs with 'predefined' causes, and for that reason could not be included in this study. This is an approach sometimes used to analyse, for example, electrical systems. The remaining analysis covered the design areas, Powertrain (25%), Chassis (approximately 10%) and Body and Trim (approximately 65%) engineering.

The FMEA study included analyses performed both in-house (2/3) and by suppliers (1/3). For the in-house analyses, half of them were D-FMEAs and the remaining analyses were S-FMEAs. The supplier analyses were all D-FMEAs of Body and Trim subsystems or components and were all suitable for the study.

The approach in the study was to assign one of the five noise factor categories to each identified potential cause of a failure mode described in the FMEA. This was repeated for all 'causes' in all FMEAs in the study. The result was discussed and reviewed by peers with long experience of quality and reliability engineering, a review that confirmed the result.

To be able to clearly understand the nature of the potential risks identified in the analysis studied, it was often necessary to read the whole description of the potential failure mechanism.

4.4 Result

The result of matching potential causes with the five noise factor categories is shown below. Causes that did not match any of the five categories are collected in a 'non-noise factor' category and the result of the matching process is discussed below.

4.4.1 Causes Matching the Noise Factor Categories

The study showed that it was possible to assign one of the noise factor categories to a *majority* of the causes that were identified in the FMEAs. Some examples in each noise factor category are:

1. 'Piece-to-piece variation' included causes of failure due to tolerances, tolerance chain, assembly, production variation, manufacturing process or concerns regarding transport and handling.
2. In the category 'Changes over time', the causes were described as fatigue, wear out, weld crack, insufficient durability, relaxation or wear due to different materials.
3. 'Customer usage' noise factors/causes included excess speed, misuse, scratches, stains from cleaning or washer fluid freezing.
4. The category 'External environment' included causes such as clogging, thermal load, corrosion because of wet environment, moist environment causing adhesion between components, collecting of dirt, contamination, salt deposits, high friction due to cold/hot conditions or the presence of foreign particles.
5. 'Systems interaction' causes included overheating, foreign particles from nearby systems, corrosion through interaction with nearby systems or interference from cables and other components.

It should be said about the noise factor categorization that it was fairly easy to assign a category to a potential cause, although the categories sometimes 'overlap'. An example of where it was more difficult to categorize was the case of a component experiencing excessive 'inner wear' due to production variations. In this case it was classified as piece-to-piece variation, since not all parts of the type would experience that wear. Another example was a potential failure with a component that could corrode due to a salty environment. In that case the cause was classified as 'external environment' as without that environment there would be no corrosion. A third example is a component that could experience a high external temperature load, and the question was whether it was caused by heat generated from neighbouring systems (systems interaction) or a hot external environment. From the analysis it was clear what the potential cause was and a category could be selected to reflect that.

4.4.2 Causes not Assignable to any of the Noise Factor Categories

It turned out that about 30% of the potential causes *could not* be classified by the five noise factor categories as they just did not fit into any of the categories. This result first seemed to be contradictory to the original assumption that causes in FMEAs and noise factors are actually the same phenomena, only expressed in a different way.

In order to understand the nature of the causes that did not fit into the noise factor categories, here are some examples: not sufficient function, do not meet serviceability requirements, do not meet assembly requirements, poor appearance due to tooling error or other specification issues. A common theme in this category of causes seems to be a concern about the possibility that the 'nominal' design does not meet the requested function, *regardless* of whether the product is affected by a noise factor or not. A product function of this kind is one that can be requested by an external as well as an internal customer, e.g. producability requested by manufacturing.

To label the additional category, the three-step procedure for countermeasures by design, i.e. system design, parameter design and tolerance design, outlined in the Off-Line Quality Control concept (Taguchi and Wu, 1979) was used. According to this approach, the result of the system design phase is a product that, if not affected by noise factors, meets customer expectations. It is during the parameter design phase that the proposed product design should be 'optimized', which includes minimized noise factor influence on the product. In other words, the product must be robust against the influence of the noise factors. The last step, tolerance design, is aimed at defining tolerance limits ('allowance limits') for minimal quality loss.

The causes/issues in the additional category are more connected to the 'system design' phase, using Taguchi's nomenclature. Consequently, this additional category of causes was given the name 'system design' concern/causes and by adding that sixth category, a category of causes not connected to noise factor categories, it *was possible* to place all causes in the FMEAs studied into one of the five + one categories (Figure 4.3).

4.4.3 Comments on the Result

The major 'noise/cause categories' in this evaluation are 'system design' causes and 'piece-to-piece variation'. It is also noticeable that the suppliers, in the analysis studied, predominate in these two categories as opposed to the other four. This can be the result of the specific designs that were analysed in this study, or, more likely, reflect the supplier's situation close to a production process and their efforts to fulfill the customer's specifications. 'Customer

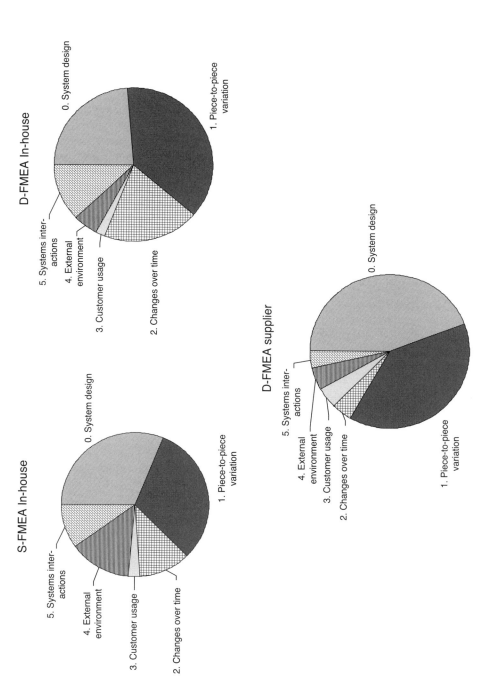

Figure 4.3 Distribution of causes into the five noise categories and the additional 'System design' category for the different types of design Failure Mode and Effect Analysis (FMEA) that were evaluated.

usage' has a low percentage in all the categories of FMEAs studied, although there was a clear difference between individual FMEAs (not shown here):

1. The first important comment that must be made is that there is not one distribution that is correct and others that are wrong; each distribution must be evaluated in the light of the analysis object. In one analysis it may be correct to have few instances of 'customer usage' whereas in an other analysis that noise factor category may occur frequently. The same is true for other categories such as 'piece-to-piece variation'.
2. The second point is that it may be reasonable that 'system design' causes have a higher percentage in earlier stages (S-FMEAs) than later (D-FMEAs), as the detailed design is less complete in the earlier phases, resulting in a higher uncertainty of requirement fulfillment.
3. The diagrams do not show the importance of different cause/noise categories, only the frequency by which they occur in the FMEAs. Frequency and importance may differ.
4. A more in-depth study of the individual FMEAs is necessary for an assessment of the relevance of the identified noise factors.

4.5 Discussion and Further Research

How can it be judged whether a FMEA is done in a proper way? What does 'proper way' mean? These issues are discussed in some texts on the FMEA method (e.g. Britsman et al., 1993; ASQC/AIAG, 2001). By adding an evaluation of noise factors, similar to that in this study, an opportunity to discuss and evaluate whether relevant noise factors are handled in an individual FMEA is provided. Engineering knowledge and experience as well as results from other analyses (e.g. parameter diagram) can provide guidance on what noise factors to be considered. In many companies, development work is done both in-house and in collaboration with suppliers, and an improved FMEA evaluation can contribute to an open dialogue and an improved application of the FMEA method.

A disadvantage of modifying a well-known method such as FMEA can be that, although the basic approach is known by engineers, the procedure differs from what they are used to and this can create some initial resistance to the modified method. So if this concept of connecting causes and noise factors is to be implemented in the FMEA procedure, how can that be done with a minimum of modification of the form? One minor modification is just to add another narrow column where the cause/noise category is specified for each potential failure mode. An added advantage of this is that the readability of the analysis will improve as it is clear for the reader what type of interaction that can cause the failure mode. Instead of introducing a new column it is possible to describe the noise/cause category by, for example, a symbol or a number in the existing 'cause of failure' column. Another way to adopt this cause/noise idea is to perform the FMEAs as it is done today but adding this concept for evaluating analyses already completed, similar to what is shown in this study. This will help engineers to quickly review the FMEA and evaluate if known noise factors are covered.

A more indirect benefit of applying the cause/noise concept in FMEAs is that it supports the general awareness of noise factors since the FMEA is a frequently used method in the industry (e.g. Arvidsson and Gremyr, 2003). By increasing the awareness of variation and noise factors, the potential for robustness thinking to have an impact on research and development engineering will increase. Thus, by integrating a frequently used reliability method

(FMEA) and Robust Design Methodology principles, a synergy effect may occur that enables a development process to deliver products and processes with less potential failures.

In this study only D-FMEAs were analysed. Can the same concept of noise/cause categories be applied in P-FMEAs and/or even be expanded to other industries?

4.6 Summary

The case study shows that it was possible to connect the majority of 'causes' in FMEAs with categories of noise factors, but an extra 'cause-category' had to be added for causes that were related to a products nominal performance, independent of noise factor interaction.

The Failure Mode and Effect Analysis form can remain unchanged or it can be just slightly modified and still enable noise factor identification in the analysis. Thus the form and method remain the same for users. A minor modification of the method may encourage the FMEA teams to consider noise factors during the analysis.

Robustness thinking and noise factor awareness can be reinforced by introducing noise factor categories in already well-established reliability methods – FMEA being one example. Efforts like this may support integration of Reliability Engineering and Robust Design Methodology.

References

Andersson, P. A practical method for noise identification and failure mode importance ranking in robust design engineering. In *III International Congress of Project Engineering*, Barcelona, 1996.

Arvidsson, M. and Gremyr, I. Robust design methodology: Status in the Swedish manufacturing industry. *Quality and Reliability International*, **19**: 285–293, 2003.

Arvidsson, M. and Gremyr, I. A review of robust design. In *8th QMOD Conference*, Palermo, 29 June–1 July, 2005.

ASQC/AIAG. *ASQC/AIAG Potential Failure Mode and Effect Analysis (FMEA)*, 3rd edn, American Society of Quality Control and Automotive Industry Action Group, Southfield, MI, July 2001.

Bergman, B. *Industriell försöksplanering och robust konstruktion*. Studentlitteratur, Lund, 1992.

Britsman, C., Lönnqvist, Å. and Ottosson, S. O. *Handbok i FMEA, Failure Mode and Effect Analysis*. Industrilitteratur, Stockholm, 1993.

Chakhunashvili, A., Johansson, P. and Bergman, B. Variation mode and effect analysis. *Annual Reliability and Maintainability Symposium*, Los Angeles, 2004.

Clausing, D. *Total Quality Development – A Step-By-Step Guide to World-Class Concurrent Engineering*. ASME Press, New York, 1994.

Davis, T. P. Science, engineering, and statistics. *ENBIS International Conference*, Copenhagen, 2004.

IEC60300-3-1. *International Standard IEC 60300-3-1, Dependability Management Part 3-1: Application Guide Analysis Techniques for Dependability Guide on Methodology*, 2nd edn, International Electrotechnical Commission, Geneva, 01, 2003.

Lönnqvist, Å. Design for Six Sigma roadmaps. *Ninth QMOD*, Liverpool, 8–10 August, 2006.

Taguchi, G. and Wu, Y. *Introduction to Off-Line Quality Control*. Central Japan Quality Control Association, Nagoya, 1979.

5

Robust Product Development Using Variation Mode and Effect Analysis

Alexander Chakhunashvili, Stefano Barone, Per Johansson and Bo Bergman

5.1 Introduction

It is a well-known fact in business and industry that unwanted variation in product characteristics is a major cause of high costs and customer dissatisfaction. Henceforth, the term product will be used to indicate a physical unit, an intangible service or a production process, of interest to an engineer in different development phases. To consistently work on issues related to variation, organizations are increasingly stimulating the implementation of Six Sigma programmes (see e.g. Harry, 1998; Magnusson et al., 2003; Park, 2003). Six Sigma deploys statistical thinking, methods and tools throughout the organization, aimed at analysing and reducing unwanted variation to the smallest possible level. While Six Sigma focuses on managing variation in manufacturing processes, some authors suggest the use of Design for Six Sigma (DFSS) in product development (see Mader, 2002; Tennant, 2002). Six Sigma and DFSS programmes utilize Statistical Process Control (SPC) to monitor variation during manufacturing processes and Robust Design Methodology (RDM) to make products insensitive to noise factors arising in production (unit-to-unit variation) or during the usage of the product (environmental/operating conditions; wear/degradation). Traditionally, Failure Mode and Effect Analysis (FMEA) (see e.g. Kececioglu, 1991; Stamatis, 1994) is also included in Six Sigma and DFSS programmes. Engineers use FMEA to identify potential failure modes and find their causes and to determine the effects on the customer. However, failures are often caused by variation. This can be shown by the well-known Load–Strength scheme

Robust Design Methodology for Reliability: Exploring the Effects of Variation and Uncertainty
edited by B. Bergman, J. de Maré, S. Lorén, T. Svensson
© 2009, John Wiley & Sons, Ltd

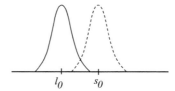

Figure 5.1   Load–Strength scheme.

(see O'Connor, 1989), illustrated in Figure 5.1. If the nominal strength (s_0) is higher than the nominal load (l_0), no failures should supposedly occur. However, either load or strength or both of them are generally affected by random variation about their nominal values. Consequently, failures might occur when a particularly high load 'meets' a particularly low strength.

This scheme, usually adopted for assessing the reliability of mechanical systems, is conceptually extendable to other situations. Some examples include:

- the number of concomitant telephone calls (load) to a call centre with a fixed number of operators (strength)
- the orders (load) arriving at a production plant at a fixed capacity (strength)
- the number of simultaneously running tasks (load) in a computer operating system with a fixed available memory (strength).

As most failures are caused by unwanted variation, it is advisable to focus directly on variation as early as possible in product development. Introducing the concept of variation and addressing the related risks in product development is not new in the literature (see Taguchi, 1986; Phadke, 1989; Clausing, 1994; Thornton, 2004). However, what has been missing until now is a practical tool used on a systematic basis to identify, assess and manage unwanted variation. Variation Mode and Effect Analysis (VMEA) is developed to fill this gap by giving engineers a quality improvement method that is applicable throughout the product development process. The purpose of this chapter is to extend the ideas proposed in Johansson et al. (2006), where a thorough presentation of a basic VMEA method was provided together with an industrial application. This extension is aimed at providing engineers with more refined techniques to be utilized in more advanced product development phases.

This chapter has the following structure. Section 5.2 gives a general overview of the VMEA method. Section 5.3 reviews the *basic* VMEA presented in Johansson et al. (2006). Section 5.4 and 5.5 present the *enhanced* and the *probabilistic* VMEAs. Section 5.6 provides an illustrative example showing when and how the three procedures can be applied. Section 5.7 summarizes the main contributions and gives final remarks.

5.2 Overview of the VMEA Method

The VMEA method starts with the identification of critical product characteristics, henceforth called Key Product Characteristics (KPC). They are characteristics that require particular attention because their excessive variation might adversely affect product functions, safety, compliance with imposed regulations and, more generally, the quality of the product. The concept of KPC and its definition are well known in industry. ISO/TS16949 (2002) uses the term special characteristics and defines them as 'product characteristics or manufacturing process parameters subject to variation which may affect safety or compliance with regulations, fit, function, performance or subsequent processing of product'. The KPCs can be obtained from a previously performed Quality Function Deployment (QFD) study (see Hauser and Clausing, 1988; Cohen, 1995). Alternatively, a cross-functional team of engineers can propose them as a result of a brainstorming session.

To understand how variation is transferred, the selected KPC is broken down into a number of sub-KPCs, defined as known and controllable subcharacteristics that influence the KPC. For each sub-KPC, one or several noise factors (NFs) are identified. The NFs are sources of variation that cause deviations in sub-KPCs. They usually stem from different production and usage situations as a result of variability in manufacturing processes, product deterioration and influence of environmental factors (see e.g. Kackar, 1985).

While the process described above is a two-level KPC causal breakdown, it can be generalized at any chosen level. It constitutes a basis for making sensitivity and variation size assessments, which in turn allows calculation of a Variation Risk Priority Number (VRPN), the main variation risk indicator directing engineers' attention to the areas of the product to which excessive variation might be detrimental.

Depending on the product development phase, engineers can carry out a *basic*, an *enhanced* or a *probabilistic* VMEA. While the structure of the procedure is the same for all three VMEAs, the assessment methods differ somewhat from each other. In the early phases of development, when only limited information is available, a *basic* VMEA is the natural choice. The *basic* VMEA is easy to follow. The sensitivity and variation size assessments are subjective and are usually made on a 1–10 scale. The idea is to quickly learn about potential sources of variation and assess their impact on KPCs. As progress is made through the initial phases of development, more information becomes available. It is then possible for engineers to make an *enhanced* VMEA by using more accurate sensitivity and variation size assessments. At this stage, engineers can verify the assumptions they have made with regard to sub-KPCs and NFs. Since the enhanced VMEA generally yields more accurate results it is worthwhile to verify that the underlying KPC, sub-KPC and NF model adequately reflects reality. Finally, as the development enters later phases and some quantitative data on test results or other measurements become available, engineers can apply a *probabilistic* VMEA. The probabilistic VMEA gives, in most cases, the most accurate estimation of the sensitivities and variation size, thus resulting in the most accurate VRPNs. An illustration of the application of VMEA throughout the product development process is given in Figure 5.2.

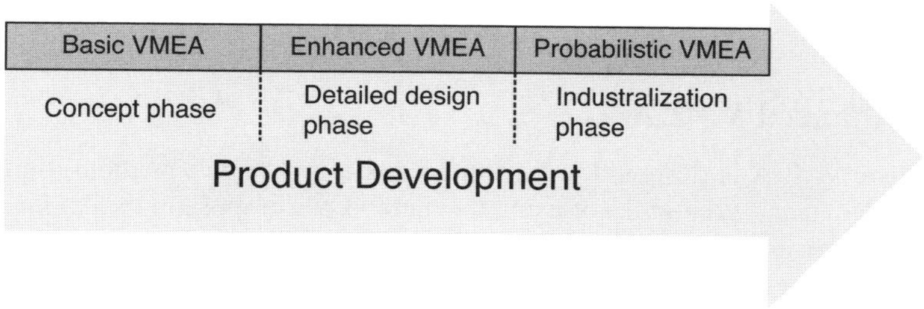

Basic VMEA	Enhanced VMEA	Probabilistic VMEA
Concept phase	Detailed design phase	Industralization phase

Product Development

Figure 5.2 Three variation mode and effect analysis (VMEA) procedures applied in product development.

5.2.1 A General Procedure for VMEA

The general procedure for making a VMEA is the same for all three VMEA types. We briefly describe this general four-step procedure below:

1. *KPC causal breakdown* – in the first step, the KPC is decomposed into a number of sub-KPCs and NFs affecting the sub-KPCs.
2. *Sensitivity assessment* – in the second step, engineers assess the sensitivity of the KPC to the action of each sub-KPC as well as the sensitivity of the sub-KPCs to the action of each NF.
3. *Variation size assessment* – in the third step, engineers examine NFs and assess their variation size.
4. *Variation Risk Assessment and Prioritization* – in the fourth step, a VRPN is calculated for each NF based on the assessments made in the previous three steps:

$$VRPN_{NF} = \alpha_i^2 \alpha_{ij}^2 \sigma_{ij}^2, \tag{5.1}$$

where α_i indicates the sensitivity of the KPC to the action of the ith Sub-KPC ($i = 1, 2, \cdots, m$), α_{ij} indicates the sensitivity of the ith sub-KPC to the action of the jth NF, and σ_{ij} indicates its variation size. If a sub-KPC is influenced by several NFs, it is possible to calculate a VRPN relative to that sub-KPC as a sum of the VRPNs for each NF acting on the selected sub-KPC:

$$VRPN_{Sub-KPC} = \sum VRPN_{NF}. \tag{5.2}$$

These VRPNs provide a basis for a Pareto diagram, resulting in the prioritization of sub-KPCs and NFs based on their contribution to the total variability of the KPC. The Pareto diagram is important for selecting the product area(s) that require most attention. This might imply actions aimed at eliminating some of the sources of variation or increasing the robustness of the product characteristics to sources of variation. The formal justification of the method is provided in the Appendix.

5.3 The Basic VMEA

In the basic VMEA, the assessments of sensitivities and variation size are made on a scale of 1–10, making the application process plain and simple. To conduct an adequate VMEA that incorporates different views and competences, a cross-functional team of engineers should be formed. The criteria for the sensitivity and variation size assessment are given in Tables 5.1 and 5.2, respectively.

5.4 The Enhanced VMEA

While the *basic* VMEA is designed to be simple and easy to apply without a great deal of previous knowledge and experience of using the method, we propose in the *enhanced* VMEA a number of enhancements that ensure a more accurate outcome. One such enhancement concerns the assessment method, i.e. the assessment of sensitivities and the assessment of NF variation. Like the *basic* VMEA, the *enhanced* VMEA also employs the four-step procedure described in Section 5.2.1.

Table 5.1 Sensitivity assessment criteria.

Category	Criteria for assessing the sensitivity	Weight
Ver low	The variation of NF (sub-KPC) is (almost) not at all transmitted to sub-KPC (KPC)	1–2
Low	The variation of NF (sub-KPC) is transmitted to sub-KPC (KPC) to a small degree	3–4
Moderate	The variation of NF (sub-KPC) is transmitted to sub-KPC (KPC) to a moderate degree	5–6
High	The variation of NF (sub-KPC) is transmitted to sub-KPC (KPC) to a high degree	7–8
Very high	The variation of NF (sub-KPC) is transmitted to sub-KPC (KPC) to a very high degree	9–10

5.4.1 Assessment of sensitivities

In order to better understand the enhanced assessment scale, let us consider a relationship between a KPC and a generic sub-KPC , analytically described by a function $Y = \varphi(X)$. The sensitivity of Y to X is the first derivative of the function and it is graphically represented by the slope of the curve, as illustrated in Figure 5.3.

Obviously, the sensitivity will depend on the point, μ_x, at which it is calculated. The sensitivity determines how much of the variation of a sub-KPC is transferred to the KPC (the same reasoning can be applied by replacing the KPC with a sub-KPC and the sub-KPC with a NF). The absolute value of the sensitivity can theoretically range between 0 and $+\infty$. It is possible to graphically represent a sensitivity value by means of a line drawn in the positive quadrant of the coordinate system. A horizontal line represents the sensitivity value equal to 0, a vertical line represents the sensitivity value equal to $+\infty$, and a line drawn at 45° represents a sensitivity value equal to 1.

A new sensitivity assessment scale, called *sensitivity fan*, spans the positive quadrant of the Cartesian graph, dividing it into 11 sectors of approximately the same angular size; see Figure 5.4. The idea of assessing sensitivities using this assessment instead of a traditional 1–10 scale comes from the fact that it takes into account the real nature of sensitivity. Each of the lines on the sensitivity fan corresponds to a specific sensitivity value. For practical reasons, the points on the scale are chosen to make the assessment as easy as possible. The last scale line is not vertical because it would in that case correspond to an unrealistic scale value equal to $+\infty$. The scale also contains values 0 and 1, representing *absolute insensitivity*

Table 5.2 Noise factor (NF) variation assessment criteria.

Category	Criteria for assessing the variation of NF	Weight
Very low	NF is considered to be almost constant	1–2
Low	NF exhibits small fluctuations	3–4
Moderate	NF exhibits moderate fluctuations	5–6
High	NF exhibits visible and high fluctuations	7–8
Very high	NF exhibits very high fluctuations	9–10

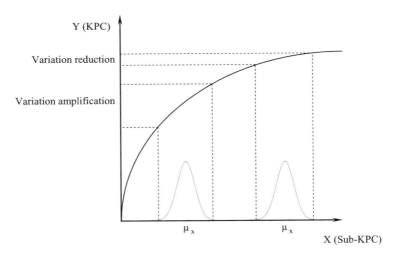

Figure 5.3   Variation transmission scheme: amplification/reduction.

and *direct proportional (additive) transfer* cases, respectively. The scale value 0 can be used when it is believed that variation in a sub-KPC is not transferred to the KPC. The scale value 1 can be used when it is believed that a sub-KPC transfers as much variation to the KPC as it exhibits. In other words, there is neither amplification nor reduction of variation in the transfer process. Consequently, a sensitivity value equal to unity corresponds to a neutral position. In general, we can say that sensitivity values less than unity (low–moderate sensitivities) represent cases of variation reduction or variation smoothing, while sensitivity values higher than unity (high–highest sensitivities) represent cases of variation amplification. Although the latter may be rather common in the initial phases of design, the highest sensitivity values should

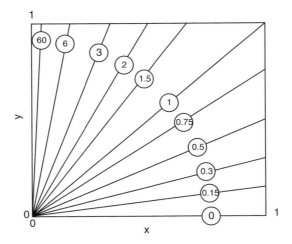

Figure 5.4   The sensitivity fan, a new 11-point scale for sensitivity assessment.

gradually decrease to minimum, leading to a greater degree of robustness in the given design. Furthermore, the sensitivity values greater than unity should be taken as early indications of an inconsistent design and in that case action aimed at decreasing them should be undertaken immediately.

5.4.2 Assessment of variation size

Similar to the new sensitivity assessment scale, it is possible to define a method that will ensure a more accurate assessment of NF variation size. This seems to be easily accomplishable as engineers can often quantify the NF variation size by means of variation range. For instance, engineers commonly say 'the outside temperature is a noise factor. It can range between 15 and 40°C'. In such a case, i.e. when it is possible to provide the estimates of a range, one way to assess the NF variation size is to divide the given range by six (see e.g. Montgomery, 2001). If we denote range by R, we can assess the size of the NF variation with the following formula:

$$\hat{\sigma} = \frac{R}{6}.$$

After assessing sensitivities and NF variation sizes, the VRPNs are calculated in the same way as shown in the basic VMEA, according to Equations (5.1) and (5.2).

5.5 The Probabilistic VMEA

While the *basic* and *enhanced* VMEAs can be carried and with limited information, completing the probabilistic VMEA requires that either analytical or numerical relationships between KPC, sub-KPCs and NFs are known. The *probabilistic* VMEA, expanding on the theoretical basis of the VMEA method, i.e. the Method of Moments (MOM, see the Appendix), provides a near perfect approximation of sensitivities and variation size. Thus, a VMEA performed using the MOM makes the best use of the variation transfer model from which the VRPNs are calculated. However, analytical and/or numerical relationships between KPC, sub-KPCs and NFs are not known until late phases of product development. Consequently, although the *probabilistic* VMEA provides good accuracy, it is primarily useful for verifying the previously obtained results.

5.6 An Illustrative Example

To show when and how the three VMEA procedures presented above can be applied, let us consider an illustrative example concerning the manufacturing of a rectangular flat metal plate. The plates are manufactured by cutting a metal sheet into four pieces. See the schematic drawing in Figure 5.5.

The cuttings are made using two different machines of the same model. However, the machines are not entirely precise in following the ideal cutting line; both are affected by sources of variation. Hence, deviations are observed around the nominal values of the plate length, $\mu_1 = 500$ mm and the plate width, $\mu_2 = 400$ mm.

Suppose engineers are interested in the diagonal of the plate. Thus, they regard it as the KPC. It is evident that the variability of the plate length and width will lead to variation in the

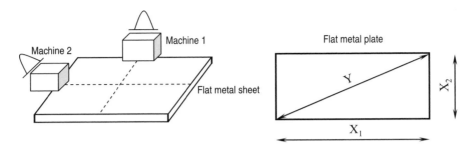

Figure 5.5   Rectangular flat metal plate.

diagonal. As engineers become aware of this quality problem, they want to analyse it in order to improve the cutting operation.

5.6.1 Application of the Basic VMEA

Although no data were initially available regarding the metal sheet cutting process, engineers wanted to exploit their knowledge and experience, and they made an analysis using the *basic* VMEA. To accomplish this task, they first broke down the KPC (diagonal Y) into a number of sub-KPCs. For the sake of simplicity, they assumed that the KPC was affected by only two sub-KPCs, i.e. by the length (X_1) and width (X_2) of the plate. Each has a nominal value around which variation is observed due to the imprecision of the cutting machines. The causal breakdown is illustrated in Figure 5.6.

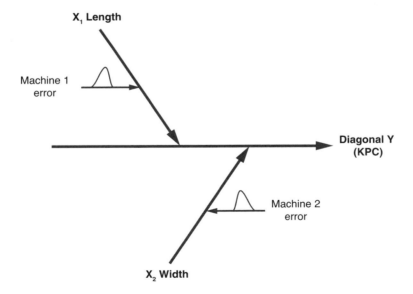

Figure 5.6   Breakdown of the flat plate diagonal.

Table 5.3 Results of the basic VMEA.

KPC	Sub-KPC	KPC sensitivity to sub-KPC	Noise factor	Sub-KPC sensitivity to NF	NF variation size	VRPN (NF)	VRPN (sub-KPC)
Diagonal, Y	Length X_1	4	N_{11}	5	8	25,600	25,600
	Width X_2	3	N_{21}	5	7	11,025	11,025

In the second step, engineers assessed the sensitivity of the diagonal Y (KPC) to the length X_1 (sub-KPC$_1$) and width X_2 (sub-KPC$_2$) on a scale of 1–10. Engineers, on the basis of their past experience, agreed that sensitivity of Y to X_1 should be higher than sensitivity of Y to X_2. They also agreed that both sensitivities were relatively low. Therefore, they placed weights 4 and 3, respectively. Next, they assessed the sensitivity of X_1 to the NF 'Machine 1 error' and the sensitivity of X_2 to the NF 'Machine 2 error'. Engineers thought that the impact of the NFs on both sub-KPCs was equal because the manufacturing process is the same and the machines are of the same model. They also believed that the weight was moderate. Therefore, they placed an equal weight of 5 on both of them. In the third step, engineers assessed the variation size of the two NFs. They knew that the cutting machine acting on X_1 was older than the cutting machine acting on X_2. Consequently, they accepted that the size of the variation of 'Machine 1 error' was slightly larger than that of 'Machine 2 error'. Furthermore, they believed that this variation was relatively high and thus placed weights 8 and 7, respectively. The final results of the *basic* VMEA are summarized in Table 5.3. As we can see from the calculations of VRPN, the estimated portion of variation contributed by the variation of X_1 is 2.3 times greater than the portion of variation contributed by X_2.

5.6.2 Application of the Enhanced VMEA

Some time later, engineers made an *enhanced* VMEA. They came together and made new sensitivity and variation size assessments. The engineers used the 11-point sensitivity fan for the sensitivity assessment and the range method as outlined in Section 5.4 for the variation size assessment. The results of the *enhanced* VMEA are given in Table 5.4. As can be seen from the calculations of VRPN, the estimated portion of variation contributed by the variation of X_1 is 3.5 times greater than the portion of variation contributed by the variation of X_2.

Table 5.4 Results of the enhanced VMEA.

KPC	Sub-KPC	KPC sensitivity to sub-KPC	Noise factor	Sub-KPC sensitivity to NF	NF variation size	VRPN (NF)	VRPN (sub-KPC)
Diagonal, Y	Length X_1	0.75	N_{11}	1	0.75	0.316	0.316
	Width X_2	0.50	N_{21}	1	0.60	0.090	0.090

5.6.3 Application of the probabilistic VMEA

The relationship between the length, width and the diagonal of the plate is defined as:

$$Y = \sqrt{X_1^2 + X_2^2}.$$

According to Equation (5.4) in the Appendix, we obtain the sensitivity coefficients:

$$\alpha_1 = \frac{\mu_1}{\sqrt{\mu_1^2 + \mu_2^2}}, \qquad \alpha_2 = \frac{\mu_2}{\sqrt{\mu_1^2 + \mu_2^2}}.$$

Variation in length and width can be modelled by:

$$X_1 = \mu_1 + N_{11}, \quad \text{and} \quad X_2 = \mu_2 + N_{21},$$

where N_{11} and N_{21} are noise factors causing variation in the cutting processes. As the cutting process involves two different machines, we can assume that N_{11} and N_{21} are independent random variables with:

$$E[N_{11}] = 0, \quad E[N_{21}] = 0, \quad Var(N_{11}) = \sigma_{11}^2, \quad Var(N_{21}) = \sigma_{21}^2.$$

By applying Equation (5.5) in the Appendix to this case, we obtain:

$$\sigma_y^2 = \frac{\mu_1^2}{\mu_1^2 + \mu_2^2}\sigma_{11}^2 + \frac{\mu_2^2}{\mu_1^2 + \mu_2^2}\sigma_{21}^2 = \frac{\mu_1^2\sigma_{11}^2 + \mu_2^2\sigma_{21}^2}{\mu_1^2 + \mu_2^2}. \qquad (5.3)$$

If we assume that the variances of the two cutting processes are equal (e.g. if the machines are of the same model and they have same maintenance schedule) and denote their common value by σ_e^2, from Equation (5.3) we obtain the nonintuitive result $\sigma_y^2 = \sigma_e^2$.

To apply the *probabilistic* VMEA, engineers needed to collect data. In this example, the nominal length was $\mu_1 = 500$ mm and the nominal width was $\mu_2 = 400$ mm, as well as $\sigma_{11} = 0.70$ and $\sigma_{21} = 0.50$. According to Equation (5.6) in the Appendix we obtain: $\alpha_1 = 0.78$ and $\alpha_2 = 0.62$. Table 5.5 provides the results of the *probabilistic* VMEA. As we can see from the calculations of VRPN, the estimated portion of variation contributed by X_1 is 3.06 times greater than the portion of variation contributed by X_2.

As a summary, Figure 5.7 illustrates the results obtained from *basic*, *enhanced* and *probabilistic* VMEAs.

As can be seen, in all three VMEAs the length, X_1, is a more critical characteristic from the standpoint of variation than the width, X_2. Consequently, the actions aimed at managing variation in the metal sheet cutting process should first be directed to the length, X_1.

Table 5.5   Results of the probabilistic VMEA.

KPC	Sub-KPC	Partial derivative	Noise factor	Sub-KPC sensitivity to NF	Sigma	VRPN (NF)	VRPN (sub-KPC)
Diagonal, Y	Length X_1	0.78	N_{11}	1	0.70	0.299	0.299
	Width X_2	0.62	N_{21}	1	0.50	0.098	0.098

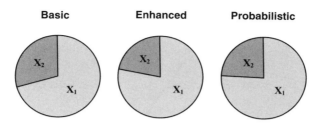

Figure 5.7   Results of application of the three VMEA procedures.

5.7 Discussion and Concluding Remarks

Despite our efforts to address the problems related to variation early in product development, we are not always successful. Therefore, we need to compensate for the shortcomings in later phases. There are many reasons for this; one is that we do not have practical methods for managing variation over the entire development process. At initial phases of development, we often face situations where the transfer function, i.e. the relationship between the KPC and the factors affecting it, is unknown. Thus, it is not possible to apply quantitative methods in order to model this relationship. Nevertheless, it is important for us to know the critical areas from a variation viewpoint. In other words, we want to know which product characteristics transmit the greatest part of the variation and what are the sources of variation affecting them. This chapter argues and shows that in this kind of situation, VMEA, especially in its basic form, can be a useful tool for extracting and quantifying information from engineering knowledge and experience. Although assessments made in the *basic* VMEA are subjective, the outcome in terms of VRPNs provides a basis for managing variation and hence improving robustness. Furthermore, the *enhanced* and *probabilistic* VMEAs provide better estimates of the portion of variation transmitted from the subcharacteristics to the response. However, we should bear in mind that they might require more quantitative data for sensitivity and variation size calculations.

In developing the VMEA procedure it has been assumed that sub-KPCs are independent (see Appendix). This assumption is reasonable if engineers have made a good choice of sub-KPCs and NFs. In fact, if there is a correlation or any other functional relationship between, for example, two sub-KPCs, one of them can obviously be excluded from the model.

The second assumption made in the VMEA method is the negligibility of interaction effects, i.e. the concomitant effect of two factors on the response of interest, KPC. These interaction effects are analytically modelled by the second-order mixed derivative terms expressed in Equation (5.5) of the Appendix. This assumption gives rise to a simple and manageable formulation and applicability of the VMEA method. The interested reader can refer to the analytical results obtained when these terms are not neglected; see for example Evans (1974, 1975). This assumption, which seems rather strong, is also motivated by two facts: (1) the FMEA method, which is widespread in the industrial context, is also based on a first-order model; (2) what generally happens in real problems is the so-called 'principle of natural hierarchical effects ordering' (see Box et al., 1978). It states that main effects (first-order effects) usually have more relevance than the second-order effects and so on. Conversely, when it is, by a priori knowledge, believed that interaction effects can be relevant, it is possible

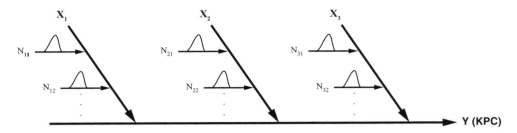

Figure 5.8   A two-level KPC causal breakdown.

to test this assumption by making some screening experiments. If, based on the experiment results, the interaction effects prove to be significant, methods other than VMEA should be utilized.

While addressing problems related to unwanted variation in products, processes or services is important, the greatest source of variation is often found to lie with the customer. Customer variation can be seen both when studying and assessing customer wants and when analysing a large number of possible ways in which the customers might use the product. Consequently, further development of VMEA should focus on incorporating support for analysing and assessing variation of the customer similar to that in Chapters 6 and 7.

Appendix: Formal Justification of the VMEA Method

This appendix gives the formal justification of the VMEA method in the case of a two-level KPC breakdown. These formulations can easily be extended to an any-level breakdown case.

Denote the KPC by Y (upper case italics letters indicate random variables). A generic sub-KPC affecting Y is denoted by X_i ($i = 1, 2, \ldots, m$), and a generic NF affecting X_i is denoted by N_{ij} ($i = 1, 2, \ldots, m$ and $j = 1, 2, \ldots, n_i$). The randomness of NFs is transmitted to the sub-KPCs and, through them, finally to the KPC. The KPC breakdown generates a model, which is graphically illustrated in Figure 5.8.

The model is analytically written as:

$$Y = f(X_1, \ldots, X_m).$$

We introduce the following notation:

$$\mu_y = E[Y], \quad \sigma_y^2 = Var[Y], \quad \mu_i = E[X_i], \quad \sigma_i^2 = Var[X_i], \quad \sigma_{ij}^2 = Var[N_{ij}].$$

It is assumed that $E[N_{ij}] = 0$. Furthermore it is assumed that the sub-KPCs X_i ($i = 1, 2, \ldots, m$) are mutually stochastically independent and the NFs N_{ij} ($j = 1, 2, \ldots, n_i$) are mutually stochastically independent $\forall i = 1, 2, \ldots, m$. The sensitivity of Y to X_i is analytically defined by:

$$\alpha_i = \left. \frac{\partial Y}{\partial X_i} \right|_{\underline{\mu}}, \tag{5.4}$$

where $\underline{\mu} = \mu_1, \mu_2, \ldots, \mu_m$.

In a similar way, the sensitivity of X_i to N_{ij} is defined as:

$$\alpha_{ij} = \left. \frac{\partial X_i}{\partial N_{ij}} \right|_0 ,$$

where $\underline{0}$ is the null vector of dimension n_i.

The second-order Taylor expansion of Y is given by:

$$Y \cong f(\underline{\mu}) + \sum_{i=1}^{m} \left. \frac{\partial f}{\partial X_i} \right|_{\underline{\mu}} (X_i - \mu_i) + \sum_{i=1}^{m} \left. \frac{\partial^2 f}{\partial X_i^2} \right|_{\underline{\mu}} \frac{(X_i - \mu_i)^2}{2}$$

$$+ \sum_{i=1}^{m} \sum_{k=2, k>i}^{m} \left. \frac{\partial^2 f}{\partial X_i \partial X_k} \right|_{\underline{\mu}} (X_i - \mu_i)(X_k - \mu_k). \tag{5.5}$$

By using the previous formula, it is possible to calculate the first two moments (expected value and variance) of Y. For the expected value, the full second-order approximation is generally used. For the calculation of the variance, only the first-order terms are considered in order to obtain a simpler formulation:

$$\sigma_y^2 = \sum_{i=1}^{m} \alpha_i^2 \sigma_i^2. \tag{5.6}$$

By following the same reasoning for the sub-KPC X_i , its variance, σ_i^2, can be further decomposed as follows:

$$\sigma_i^2 = \sum_{i=1}^{n_i} \alpha_{ij}^2 \sigma_{ij}^2 \tag{5.7}$$

Therefore, by combining Equations (5.1) and (5.2) we obtain:

$$\sigma_y^2 = \sum_{i=1}^{m} \alpha_i^2 \left(\sum_{i=1}^{n_i} \alpha_{ij}^2 \sigma_{ij}^2 \right), \tag{5.8}$$

which is – for a two-level KPC breakdown – the general formulation of the so-called Method of Moments (MOM), a well-established result in statistics (see Morrison, 1998).

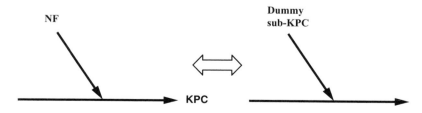

Figure 5.9 Introduction of a dummy sub-KPC.

The model described above does not exclude the possibility that an NF directly affects the KPC. In fact, in that case, it is possible to introduce a 'dummy' sub-KPC (with a sensitivity equal to 1) connecting an NF to the KPC (Figure 5.9).

References

Box, G. E. P., Hunter, S. and Hunter, W. *Statistics for Experimenters*. John Wiley & Sons, Inc., New York, 1978

Clausing, D. *Total Quality Development – A Step-By-Step Guide to World-Class Concurrent Engineering*. ASME Press, New York, 1994.

Cohen, L. *Quality Function Deployment*. Prentice Hall, Englewood Cliffs, NJ, 1995.

Evans, D. H. Statistical tolerancing: The state of the art. Part I. Background. *Journal of Quality Technology*, **6**: 188–195, 1974

Evans, D. H. Statistical Tolerancing: The state of the art. Part II. Methods for estimating moments. *Journal of Quality Technology*, **7**: 1–11, 1975

Harry, M. J. Six sigma: A breakthrough strategy for profitability. *Quality Progress*, **31**: 60–64, 1998.

Hauser, J. R. and Clausing, D. The house of quality. *The Harvard Business Review*, **May–June**, 1988.

ISO/TS16949. *Quality Management Systems: Automotive Suppliers*. International Organization for Standardization, Geneva, 2002.

Johansson, P., Chakhunashvili, A., Barone, S. and Bergman, B. Variation mode and effect analysis: a practical tool for quality improvement. *Quality and Reliability Engineering International*, **22**: 865–876, 2006.

Kackar, R. N. Off-line quality control, parameter design, and the Taguchi method (with discussion). *Journal of Quality Technology*, **17**: 176–188, 1985.

Kececioglu, D. *Reliability Engineering Handbook*. Prentice-Hall, Englewood Cliffs, NJ, 1991.

Mader, D. P. Design for six sigma. *Quality Progress*, **35**(7): 82–86, 2002.

Magnusson, K., Kroslid, D. and Bergman, B. *Six Sigma – The Pragmatic Approach*. Studentlitteratur, Lund, 2003.

Montgomery, D. C. *Introduction to Statistical Quality Control*. John Wiley & Sons, Inc., New York, 2001.

Morrison, S. J. Variance synthesis revisited. *Quality Engineering*, **11**: 149–155, 1998.

O'Connor, P. D. T. *Practical Reliability Engineering*. John Wiley & Sons Inc., New York, 1989.

Park, S. H. *Sigma for Quality and Productivity Promotion*. Asian Productivity Organization, Tokyo, 2003.

Phadke, M. S. *Quality Engineering using Robust Design*. Prentice Hall: Englewood Cliffs, NJ, 1989.

Stamatis, D. H. *Failure Mode and Effect Analysis: FMEA from Theory to Execution*. ASQ Quality Press, Milwaukee, 1994.

Taguchi, G. *Introduction to Quality Engineering – Designing Quality into Products and Processes*. Asian Productivity Organization, Tokyo, 1986.

Tennant, G. *Design for Six Sigma: Launching New Products and Services Without Failure*. Hampshire, Gower Publishing Limited, 2002.

Thornton, A. C. *Variation Risk Management: Focusing Quality Improvements in Product Development and Production*. John Wiley & Sons, Ltd, Chichester, 2004.

6

Variation Mode and Effect Analysis: An Application to Fatigue Life Prediction

Pär Johannesson, Thomas Svensson, Leif Samuelsson, Bo Bergman and Jacques de Maré

6.1 Introduction

An important goal of engineering design is to produce a reliable system, structure or component. One such well-established method is FMEA (Failure Mode and Effect Analysis), where the aim is to identify possible failure modes and evaluate their effect. A general design philosophy, within robust design, is to make designs that avoid failure modes as much as possible, see for example Davis (2006). Therefore, it is important that the design is robust against different sources of unavoidable variation. In Chapter 5 a general methodology called VMEA (Variation Mode and Effect Analysis) was presented in order to deal with this problem, see also Chakhunashvili et al. (2006). The VMEA is split into three different levels: (1) basic VMEA, in the early design stage, when we only have vague knowledge about the variation, and the goal is to compare different design concepts; (2) enhanced VMEA, further in the design process when we can better judge the sources of variation; and (3) probabilistic VMEA, in the later design stages where we have more detailed information about the structure and the sources of variation, and the goal is to be able to asses the reliability.

Here we treat the third level, the probabilistic VMEA, and we suggest a simple model, also used in Svensson (1997), for assessing the total uncertainty in a fatigue life prediction,

This chapter is based on the article by Pär Johannesson, Thomas Svensson, Leif Samuelsson, Bo Bergman and Jacques de Maré, 'Variation mode and effect analysis: An application to fatigue life prediction', *Quality and Reliability Engineering International*, Vol. 25, 2009, pp. 167–179, and is reproduced by the permission of the publisher John Wiley & Sons, Ltd.

Robust Design Methodology for Reliability: Exploring the Effects of Variation and Uncertainty
edited by B. Bergman, J. de Maré, S. Lorén, T. Svensson

Figure 6.1   Low pressure shaft in a jet engine.

where we consider different sources of variation, as well as statistical uncertainties and model uncertainties. The model may be written as

$$Y = \hat{Y} + X_1 + X_2 + \ldots + X_p + Z_1 + Z_2 + \ldots + Z_q, \qquad (6.1)$$

where X_k and Z_k are random variables representing different sources of scatter or uncertainty, respectively. In our case it is appropriate to study the logarithmic life, thus $Y = \ln N$, where N is the life. The prediction of the logarithmic life $\hat{Y} = \ln \hat{N}$ may be a complicated function, e.g. defined through finite element software; however the analysis of the prediction uncertainty is based on a linearization of the function, making use of only the sensitivity coefficients. Further, for reliability assessments the log-life, $Y = \ln N$, is approximated by a normal distribution.

The methodology will be discussed using a case study of a low pressure shaft in a jet engine (Figure 6.1). The low pressure shaft connects the low pressure turbine with the low pressure compressor. Critical points of the shaft are a number of holes used for transporting lubricant oil.

The classification of sources of prediction uncertainty is discussed in Section 6.2. In Section 6.3 the problem of life prediction is studied where a linear model of the logarithmic life is introduced and includes the different kinds of prediction uncertainties. An application of the method of estimating the size of different prediction uncertainties is given in Section 6.4. In the following section the quantification of the sum of the uncertainties is used to construct a prediction interval which is a rational way to construct safety factors. If further experiments are carried out a possibility to update the prediction interval is proposed in Section 6.6. In particular the experiments are used for eliminating a possible systematic prediction error. The last section contains concluding remarks about the applicability of the method and its connection to the first-order second moment method.

6.2 Scatter and Uncertainty

There are various ways in which the types of variation might be classified, see for example Melchers (1999), Ditlevsen and Madsen (1996), and Lodeby (2000). The first way is to distinguish between *aleatory* uncertainties and *epistemic* uncertainties. The first of these refers to the underlying, intrinsic uncertainties, e.g. the scatter in fatigue life and the load variation within a class of customers. The latter refers to the uncertainties which can be reduced by means of additional data or information, better modelling and better parameter estimation methods.

Here we will use the terminology *scatter* as being the aleatory uncertainties. In our approach, we will focus on the three kinds of uncertainties mentioned by Ditlevsen and Madsen (1996), and denote them by:

1. *Scatter* or physical uncertainty, which is uncertainty identified with the inherent random nature of the phenomenon, e.g. the variation in strength between different components.

2. *Statistical uncertainty*, which is uncertainty due to statistical estimators of physical model parameters based on available data, e.g. estimation of parameters in the Coffin–Manson model for life based on fatigue tests.
3. *Model uncertainty*, which is uncertainty associated with the use of one (or more) simplified relationship to represent the 'real' relationship or phenomenon of interest, e.g. a finite element model used for calculating stresses is only a model for the 'real' stress state.

Another important category of uncertainties is what can be called uncertainty due to human factors. These are not treated here, but must be controlled by other means, which is discussed in for example Melchers (1999). In our analysis we assume that the phenomenon is predictable. This is the case if the system and the distribution of the random variables do not change with time.

6.3 A Simple Approach to Probabilistic VMEA

Here we will present a simple model for the prediction uncertainty based on a summation of contributions from different sources of scatter and uncertainty. We will discuss it in terms of fatigue life prediction, but it can easily be adapted to other situations, e.g. prediction of maximum stress or maximum defect sizes.

6.3.1 Model for Uncertainty in Life Predictions

We will study the prediction error of the logarithmic life prediction

$$\hat{Y} = \ln \hat{N} = f(\Psi, \hat{\boldsymbol{\theta}}, \hat{\mathbf{X}}), \tag{6.2}$$

where $f(\Psi, \hat{\boldsymbol{\theta}}, \hat{\mathbf{X}})$ is our model for the life which involves the damage driving parameter, Ψ (e.g. stress, strain or force), the estimated parameter vector, $\hat{\boldsymbol{\theta}} = (\hat{\theta}_1, \ldots, \hat{\theta}_r)$, and the modelled scatter, $\hat{\mathbf{X}}$. The prediction error can be written as

$$\hat{e} = Y - \hat{Y} = \ln N - \ln \hat{N} = g(\Psi, \widetilde{\mathbf{X}}) - f(\Psi, \hat{\boldsymbol{\theta}}, \hat{\mathbf{X}}), \tag{6.3}$$

where $g(\Psi, \widetilde{\mathbf{X}})$ is the actual relation for the log-life depending on the damage driving parameter, Ψ, and the scatter involved, $\widetilde{\mathbf{X}}$. The next step is to approximate the prediction error by a sum

$$\hat{e} = X_1 + X_2 + \ldots + X_p + Z_1 + Z_2 + \ldots + Z_q, \tag{6.4}$$

where the quantities $\mathbf{X} = (X_1, \ldots, X_p)$ and $\mathbf{Z} = (Z_1, \ldots, Z_q)$ represent different types of scatters and uncertainties, respectively. The random quantities X_i and Z_j are assumed to have zero mean and variances τ_i^2 and δ_j^2, respectively. In the analysis we only use the X_i's and Z_j's variances and covariances, but not their exact distributions, which are often not known to the designer. In some situations it is more natural to estimate the scatter or uncertainty in some quantity that is related to log-life, and then use a sensitivity coefficient to obtain its effect on the log-life. One such example is the scatter in life due to geometry variations because of tolerances, where it is easier to estimate the scatter in stress, say τ_i', which is then transferred via a sensitivity coefficient to the scatter in life, $\tau_i = |c_i| \cdot \tau_i'$. This is motivated by the Gauss approximation formula for the transfer function $\ln N = h(S)$ from stress to log-life, which gives

$$\ln N = h(S) \approx h(s_0) + c_i(S - s_0) \quad \text{with} \quad c_i = \left. \frac{dh}{dS} \right|_{S=s_0}, \tag{6.5}$$

where s_0 is the stress corresponding to the nominal values. Thus, the variance for the log-life is approximated by $\tau_i^2 = c_i^2 \cdot \tau_i'^2$.

When constructing reliability measures in Section 6.5, we will make use of a normal distribution approximation on the log-life. Theoretically this can be motivated through the central limit theorem (CLT), which in its simplest form states that the sum $S_n = (X_1 + X_2 + \ldots X_n)/\sqrt{n}$ converges, as n tends to infinity, to a normal distribution with zero mean and variance σ^2, if the X_k quantities are independent and equally distributed with mean zero and finite variance σ^2. Note that the convergence is regardless of the parent distribution of X_k. The theorem has many generalizations to both dependence and unequal variances. In practice the convergence may be quite fast. If, for example, we want to compute the 1% quantile, the Central Limit Theorem approximation is often good enough with only two or three terms, if the parent distribution is not too skew, and all X_k quantities are of the same order of magnitude. The level of approximation can be studied easily in a simulation study.

In the industrial example below we will consider life prediction for low cycle fatigue, where the damage driving parameter is the strain range $\Delta\varepsilon$. The life model in this case is the Coffin–Manson equation

$$\frac{\Delta\varepsilon}{2} = \frac{\sigma_f'}{E} \cdot (2N)^b + \varepsilon_f' \cdot (2N)^c, \tag{6.6}$$

which in cases of large strain ranges can be simplified to the Basquin–Coffin–Manson equation

$$\frac{\Delta\varepsilon}{2} \approx \varepsilon_f' \cdot (2N)^c, \tag{6.7}$$

where the elastic strain part is neglected. This can be rewritten as

$$N = e^a \cdot \Delta\varepsilon^{1/c} \Leftrightarrow \ln N = a + \frac{1}{c} \ln \Delta\varepsilon \text{ with } a = -\ln 2 - \frac{1}{c} \ln(2\varepsilon_f'), \tag{6.8}$$

giving the life prediction model $\ln \hat{N} = f(\Psi, \hat{\theta}, \hat{X})$ as a linear regression model, which will be used in our case study.

The logarithmic transformation used here has an important implication in that the variation measures τ_i^2 and δ_j^2 can be interpreted as coefficients of variation for the life, i.e.

$$Var[\ln N] \approx \frac{Var[N]}{(E[N])^2}. \tag{6.9}$$

This interpretation is practical when one is forced to use engineering judgements for estimates of uncertainties, since they can be easily related to percentage uncertainty.

6.4 Estimation of Prediction Uncertainty

There are different kinds of uncertainties that need to be estimated. For scatter, the straight forward method is to make an experiment and calculate the sample standard deviation. In

Table 6.1 Table summarizing the sources of scatter and uncertainty and their contributions, in terms of standard deviation of the logarithmic life, to the total prediction uncertainty.

Type of scatter and uncertainty	Logarithmic life $\ln(N)$			
	Scatter	Uncertainty	Total	Comments
Strength scatter			0.38	
Material, within shaft	0.15			See 6.4.1
Material, between shaft	0.29			See 6.4.1
Geometry	0.20			See 6.4.1
Statistical uncertainty			0.07	
LCF curve		0.07		See 6.4.2
Model uncertainty			0.84	
LCF curve		0.05		See 6.4.3
Mean stress model		0.30		See 6.4.3
Multi- to uniaxial		0.20		See 6.4.3
Plasticity		0.72		See 6.4.3
Stress analysis		0.24		See 6.4.3
Temperature		0		See 6.4.3
Load scatter and uncertainty			0.50	
Service load, scatter	0.40			See 6.4.4
Service load, uncertainty		0.30		See 6.4.4
Total	0.55	0.90	1.05	

more complicated situations, ANOVA (Analysis of Variance) is a useful tool. However, it is not always possible or economically motivated to perform experiments. Instead informed guesses, previous designs, or engineering experience have to be used. Concerning statistical uncertainty, there are standard statistical methods such as the maximum likelihood theory for finding expressions of the variances of the estimates. In more complicated situations the bootstrap method can be applied. However, in order to use these statistical methods it is required that the original data are available, which is not always the case. A typical example is that only the estimated life curve is available, and maybe also information about the number of tests performed. In these situations it is also necessary to have a method in order to form an idea of the statistical uncertainty. For a specified model the model uncertainty is in fact a systematic error. However, if we consider the prediction situation such that we randomly choose a model from a population of models, then the systematic model error appears as a random error in the prediction. We will discuss different methods for estimating the model uncertainty, e.g. considering a random choice of models from the population of models, or considering extreme cases of models. In the following, we will provide some more details and examples on the estimation, and explain the estimated values in Table 6.1.

6.4.1 Estimation of Scatter

In our case study, the evaluation of the scatter is based on fatigue tests of three different shafts resulting in six observed lives each. The critical points of the shafts are holes for transporting lubricant oil. Each shaft has three such oil-holes and each oil-hole has two critical points. It

Figure 6.2   The critical points of the shafts: holes for transporting lubricant oil. Both crack initiation and crack propagation was monitored using two crack-growth gauges for each hole.

was decided to perform a fatigue test with the possibility to monitor both crack initiation and crack propagation. This was accomplished using crack-growth gauges applied at the holes, two for each hole (Figure 6.2). The experiment thus resulted in six observed lives for each of the three shafts (Figure 6.3).

From an ANOVA it was found that there is both a within-shaft scatter and a between-shafts scatter, estimated at $\tau_{within} = 0.15$, and $\tau_{between} = 0.29$, respectively (Figure 6.3). The within-shaft scatter originates from material scatter, whereas the between-shafts scatter is production scatter due to different batches, processing, or supplier effects.

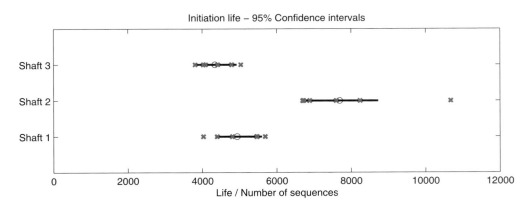

Figure 6.3   Observed lives and confidence intervals for median life of the three different shafts tested.

In many situations it is necessary to rely on engineering judgement, and for example ask a man in the workshop 'what is the worst case?'. The answer should often not be interpreted as representing zero probability of observing something more extreme, but rather as there being a very small risk of that. Therefore, the 'worst case' statement can be mathematically interpreted as a certain quantile, e.g. the 1/1000 risk of observing a more extreme case. By further assumptions, say a normal distribution, it is possible to estimate the scatter, τ, according to

$$z_{0.001} = \mu + 3\tau \Rightarrow \tau = (z_{0.001} - \mu)/3, \tag{6.10}$$

where μ is the nominal value, $z_{0.001}$ is the 'worst case' representing the 0.1% upper quantile, and the value 3 comes from the 0.1% quantile of the standard normal distribution.

The strength of a structure and a component in the structure depends not only on the material properties, but is also highly dependent on geometry and assembling quality. For the example in Table 6.1, the geometry was varied according to the tolerances and the worst case resulted in a 10% change of the calculated stress. The worst case was here interpreted as the 0.1% quantile, giving $\tau'_{geometry} = 0.10/3$. Further, the sensitivity coefficient from log-stress to log-life was estimated at $c_{geometry} = -6$, giving the estimated scatter in log-life, $\tau_{geometry} = |c_{geometry}| \cdot \tau'_{geometry} = 6 \cdot 0.10/3 = 0.20$.

6.4.2 Statistical Uncertainty

There are standard statistical methods such as the maximum likelihood theory for finding expressions of the variances of estimates, see e.g. Casella and Berger (2001) or Pawitan (2001). For more complicated estimation procedures the bootstrap method, which is a general method based on simulations, can be applied, see e.g. Efron and Tibshirani (1993), Davison and Hinkley (1997), and Hjorth (1994).

In the case of the Basquin–Coffin–Manson model (Equation 6.8), the parameters may be estimated using linear regression. In this situation the prediction uncertainty is

$$\delta = s\sqrt{1 + \frac{1}{n} + \frac{(x - \bar{x})^2}{\sum_i (x_i - \bar{x})^2}} \approx s\sqrt{1 + \frac{1}{n} + \frac{1}{n}} = s\sqrt{1 + \frac{2}{n}}, \tag{6.11}$$

where x_i is the logarithmic strain during reference test i, x is the predicted logarithmic service strain, and the last expression is a rough approximation of the root expression. This simple approximation is correct if the squared distance from the actual value x to the reference test mean level $\bar{x}$ is the same as the mean square distance in the reference tests. This roughly means that the prediction is made in the same region as the reference tests. Consequently, in the case of interpolation the approximation is usually good enough, whereas it should not be used when extrapolating outside the range of the test data.

The simplification (Equation 6.11) can be extended to models with more variables using the expression

$$\delta = s\sqrt{1 + \frac{r}{n}}, \tag{6.12}$$

where r is the number of parameters in the model. This kind of approximation is particularly useful in cases where the original data are not available, but only the estimated life curve together with the observed experimental scatter and the number of tests performed. Thus, in these situations, a rough guess of the statistical uncertainty can be obtained as $\delta = s\sqrt{r/n}$. This was used for the case study, where a four-parameter Coffin–Manson curve from the literature was used, which was based on 20 tests. Thus, the statistical uncertainty is estimated at $\delta_{stat} = 0.15\sqrt{4/20} = 0.07$, where the scatter in log-life is estimated from the fatigue tests, $s = \delta_{within} = 0.15$.

6.4.3 Model Uncertainty

Assume that we have made life predictions based on different models, and want to address the uncertainty arising from the choice of model. We will consider two situations:

1. Assume that there is one model representing the least favourable case, and another representing the most favourable case. This means that these two models represent extreme cases of models, and all other models predict lives somewhere in between.
2. Assume that the models have been arbitrarily chosen; in other words they are randomly chosen from a population of models.

In the first situation, without any other information, it is natural to assume a uniform distribution with the end-points according to the extreme cases. This gives an estimate, based on the uniform distribution, of the standard deviation

$$s_U = \frac{\ln N_{max} - \ln N_{min}}{2\sqrt{3}}, \tag{6.13}$$

with N_{min} being the shortest predicted life, and N_{max} the longest.

In the second situation, we may use the sample standard deviation as an estimator

$$s = \sqrt{\frac{1}{n-1}\sum_{k=1}^{n}(\ln N_k - \overline{\ln N})^2}, \quad \overline{\ln N} = \frac{1}{n}\sum_{k=1}^{n}\ln N_k \tag{6.14}$$

where we have used n different life models predicting the lives $N_1, N_2, \ldots, N_n$.

For the special case of two life predictions, we can compare the two approaches

$$s_U = \frac{1}{2\sqrt{3}}|\ln N_2 - \ln N_1| = 0.289|\ln N_2 - \ln N_1|, \tag{6.15}$$

$$s = \frac{1}{\sqrt{2}}|\ln N_2 - \ln N_1| = 0.707|\ln N_2 - \ln N_1|. \tag{6.16}$$

For very long load sequences (the normal situation for turbojet engine components) it is normal procedure to perform linear elastic finite element calculations. Plasticity will therefore have to be handled by plasticity models. For the plasticity model, the Linear rule and the Neuber rule for plastic correction can be seen as two extreme cases of models. In a similar situation the predictions using the two models differed by more than a factor of ten, and in that case the model uncertainty was estimated to be

$$\delta_{plasticity} = \frac{\ln N_{max} - \ln N_{min}}{2\sqrt{3}} = \frac{\ln 21 - \ln 1.77}{2\sqrt{3}} = 0.72. \tag{6.17}$$

The model errors due to mean stress correction and the conversion from multiaxial to uniaxial stress state were analysed in the same manner. Further, the stresses were calculated using finite element programs and the uncertainty in the calculated stresses were judged to be about 4%. Thus, using the sensitivity coefficient of $c_{stress} = -6$, the uncertainty in log-life is $\delta_{stress} = 6 \cdot 0.04 = 0.24$. Often in jet engine applications there are effects of high temperature. In this case this model error was judged to be negligible, hence $\delta_{temperature} = 0$.

Another approach to assess uncertainties in calculations is to rely on round-robin studies. One example is when many different engineers have been given the same calculation task. It is then often the case that all engineers present different results. The engineers can be seen as randomly chosen from the population of all engineers, and the population scatter can be estimated. In such a way uncertainties due to calculations with different methods, programs and engineers can be judged, see for example Pers et al. (1997), and Bernauer and Brocks (2002).

6.4.4 Scatter and Uncertainty in Loads

The load that a component will experience during its time in service is usually very difficult to estimate. Experience from measurements in service for certain predefined manoeuvres gives a rough estimate of the expected load scenario, but variations and extreme events should also be considered.

This problem is often hidden from the designer, since demands are already given by other departments in the company. However, it is important to estimate the load scatter and uncertainty, since this part of the load–strength problem may override other parts and make the variability in strength negligible.

In our example, the estimates are based on previous engineering experience. There is a scatter in the load due to the individual usage, which was judged to be $\tau_{load} = 0.40$. There is also an uncertainty whether the flight missions used for design really reflect the typical usage in the field, where the uncertainty was judged to be $\delta_{load} = 0.30$.

6.4.5 Total Prediction Uncertainty

The total prediction uncertainty is the sum of all the contributions; see Equation (6.4),

$$\delta_{pred} = \sqrt{\tau_1^2 + \tau_2^2 + \ldots + \tau_p^2 + \delta_1^2 + \delta_2^2 + \ldots + \delta_q^2}, \tag{6.18}$$

which is the number in the right-hand bottom corner of Table 6.1. In this example it was reasonable to assume independence between the different sources of scatter and uncertainty, hence no covariance terms appear in Equation (6.18). However, in the case of correlation between the scatters and/or uncertainties, it is simple to add these covariance terms under the root sign in Equation (6.18).

In the following section we will describe how the result can be used for life assessment. Another important use is to obtain information on where it is most efficient to try to reduce the uncertainties. In our case, the model uncertainty due to plasticity is by far the largest uncertainty, and thus it is motivated to further study the plasticity in order to reduce its model uncertainty.

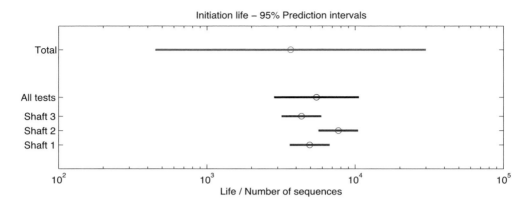

Figure 6.4   Prediction intervals for life, log-scale.

6.5 Reliability Assessment

We will now demonstrate how the estimated total prediction uncertainty in log-life can be used and presented as a prediction interval for the life or as a safety factor for the life. The analysis is based on the construction of a prediction interval using a normal approximation

$$\ln N = \ln N_{pred} \pm z_p \cdot \delta_{pred} \Rightarrow N = N_{pred} \cdot \exp(\pm z_p \cdot \delta_{pred}), \tag{6.19}$$

where N_{pred} is the life prediction according to the calculation, δ_{pred} is our previously estimated total prediction uncertainty in log-life. The factor z_p is a quantile of the standard normal distribution; $z_{0.025} = 2.0$ for a 95% interval, and $z_{0.001} = 3.0$ for a 99.8% interval.

In Figures 6.4 and 6.5 the 95% prediction interval for the calculated life predictions according to Equation (6.19) are compared to those obtained from fatigue life tests. The experiments revealed a scatter between the shafts. The experimental prediction intervals are therefore shown for each shaft individually as well as for the total scatter (within and between), compare

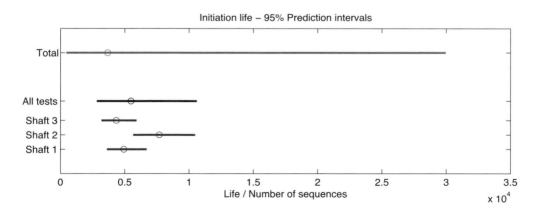

Figure 6.5   Prediction intervals for life, linear scale.

Table 6.2 Prediction quantiles for log-life.

	Quantiles				
Initiation life	0.1%	2.5%	50%	97.5%	99.9%
Life prediction	160	450	3 700	30 000	86 000

Figure 6.3. Note that the experimental life interval can be seen as a lower limit of the length of the prediction intervals, since it involves only the scatter in life that is unavoidable. The experiments could be used to update the calculation of the life prediction, i.e. to correct for the systematic error and reduce some of its uncertainties, especially the model uncertainties. This has not been pursued for this case study, but a discussion on updating is found in the next section.

For design, the lower endpoint of the prediction interval should be considered in order to obtain a reliable component. A reliability corresponding to a risk of 1/1000 is often used, corresponding to the lower endpoint of our 99.8% interval. In Table 6.2 the limits of the 95% and 99.8% prediction intervals are presented.

It is possible to define a safety factor in life, based on the prediction interval, as the ratio between the median life and a low quantile of life

$$K_p = \frac{N_{0.5}}{N_p},$$ (6.20)

where p is the probability of failure. In our case it becomes

$$K_p = \exp(\ln N_{0.5} - \ln N_p) = \exp(z_p \cdot \delta_{pred}).$$ (6.21)

For the initiation life, the safety factors in life become $K_{0.025} = 8.2$ and $K_{0.001} = 23$.

6.6 Updating the Reliability Calculation

If the total uncertainty in estimated life is dominated by uncertainties, i.e. the Z_j variables in Equation (6.4), then it is desirable to arrange validation tests to reduce these uncertainties. Assume that a test with n individual test specimens has been performed. Each individual test result can be written as

$$\ln N_i = f(\Delta\varepsilon_{tot,i}, \boldsymbol{\theta}) + \xi_i + z,$$ (6.22)

where ξ_i is the individual test specific random error, which is the sum of the different random contributions, $\xi_i = \sum_j X_{ji}$, and z is the actual model error, which is the same for all individual tests and is denoted by a lower case letter to emphasize that it is non-random. However, it may be seen as an outcome of the random distribution of model errors.

The test result can be used for estimating the total model error

$$\hat{z} = \overline{\ln N_\bullet - f(\Delta\varepsilon_{tot,\bullet}, \hat{\boldsymbol{\theta}})},$$ (6.23)

i.e. by taking the average value of the differences between the observed lives and the predictions, where we have assumed that the random errors have zero mean. The model can now be updated according to the experimental result, giving

$$\ln \hat{N} = f(\Delta \varepsilon_{tot}, \hat{\boldsymbol{\theta}}) + \hat{z}. \tag{6.24}$$

The prediction uncertainty for this updated model is now the sum of the variance of the scatter and the variance of the estimated model error $\hat{z}$. With reference to Equation (6.23) this latter property can be found from

$$Var[\hat{z}] = Var[\overline{\ln N_\bullet}] + Var[\overline{f(\Delta \varepsilon_{tot,\bullet}, \hat{\boldsymbol{\theta}})}]. \tag{6.25}$$

This is the sum originating from the statistical uncertainties in the original reference test and the new validation test. Thus, the second of these two terms comes from the statistical uncertainty in the model due to uncertain parameters, but will be decreased by averaging according to the number of individual test specimens

$$\delta_{\hat{z},\hat{\theta}}^2 = \frac{\delta_{\hat{\theta}}^2}{n}. \tag{6.26}$$

The first term comes from the random errors in the validation test and is estimated at

$$\delta_{\hat{Z},\xi}^2 = \frac{s^2}{n}, \tag{6.27}$$

where

$$s^2 = \frac{1}{n-1} \sum_{i=1}^{n} \left(\ln N_i - \overline{\ln N_\bullet} \right)^2. \tag{6.28}$$

The uncertainty in the error estimate is

$$\delta_{\hat{z}}^2 = \delta_{\hat{z},\hat{\theta}}^2 + \delta_{\hat{z},\xi}^2. \tag{6.29}$$

6.6.1 Uncertainty after Updating

In the previous section the uncertainty in model error estimate was investigated. If the validation tests are performed at conditions that are supposed to give the same model error as in service, then the uncertainty in the updated model is given by the random scatter in combination with the error uncertainty $\delta_{\hat{z}}^2$ above. If, in addition, the validation tests are performed on a random choice of components that represent all sources of scatter, then the total life uncertainty is

$$Var[\ln \hat{N}] = \delta_{\hat{z}}^2 + \tau_\xi^2. \tag{6.30}$$

However, it will often be impossible to perform validation tests under true service conditions, and this fact may add new uncertainties to the predictions. Sources for such uncertainty components must be considered, and their estimated variance must be added to the variance above. This is done in exactly the same manner as described in the Section 6.4 and results in the prediction uncertainty after updating

$$Var[\ln \hat{N}] = \delta_{\hat{z}}^2 + \tau_\xi^2 + \delta_{Z'}^2, \tag{6.31}$$

where $\delta_{Z'}^2$ is the variance of the remaining model uncertainties, coming from future service conditions that were not represented in the validation test. For instance, the loads subjected to a component in a laboratory test will not be in full agreement with the multidimensional loads acting on the component in service. Uncertainties originating from this fact must be added in service prediction.

6.7 Conclusions and Discussion

In the early design stages, the basic and enhanced VMEA should be used, and based on the results of these, the most critical components are identified. For some components it may then be motivated to make a more sophisticated probabilistic VMEA, where the scatter and uncertainties involved need to be quantified. First this can be made in quite a rough way, making use of the information available in the design process, but also making use of previous experience from similar structures, and other kinds of engineering experience. The goal is to obtain a rough estimate of the prediction uncertainty, and to locate the largest sources of scatter and uncertainty, in order to see where further efforts would be most efficient.

The proposed method represents the concept of First-Order Second-Moment (FOSM) reliability theory, see e.g. Melchers (1999) or Ditlevsen and Madsen (1996). The First-Order refers to the linearization of the objective function, and the Second-Moment refers to the fact that only the means and variances are used. The result is a prediction uncertainty in terms of the standard deviation of the logarithmic life, which can be used to calculate a prediction interval for the life or to calculate a safety factor in life.

In the present case study, the largest contribution to the prediction uncertainty originates from the model uncertainty due to the modelling of the plasticity. Therefore, in this specific case, it is motivated to further study the plasticity phenomenon. One may discuss whether the judgement of the uncertainty is realistic, and if the size of the contribution may be overestimated. However, it turns out that the model uncertainty is realistic, and it is motivated to model the plasticity phenomenon in more detail. With the computer performance of today, still increasing, it becomes more and more realistic to perform nonlinear analyses. A load sequence containing some hundred load steps is today realistic to evaluate with a nonlinear material model in the finite element calculation. The model uncertainty due to plasticity can then be expected to drop from 0.72 to 0.20, which would result in a reduction of the prediction uncertainty from 1.05 to 0.79.

Another example of the influence from tolerances was found on a similar component (also from a turbojet engine). The critical point of this component was a u-shaped groove. The influence from tolerances on the fatigue life in the u-groove was entirely dependent on the tolerances of the radius in the u-groove. A change of this radius, from minimum to maximum radius within the tolerance zone, resulted in a fourfold variation of fatigue life. A simple action in order to decrease the prediction uncertainty would here be to tighten the tolerances, which turned out to be an easy measure.

The split of the uncertainty sources into scatter and other uncertainties provides possibilities to update the reliability calculation in a rational manner when new data are available. In the case of experiments made under similar conditions as in service, the model errors may be estimated and corrected for. In the case of measurements of service loads the corresponding uncertainty entry in Table 6.1 can be updated. The updating can be useful for evaluation of field failures, for maintenance planning, and so on.

References

Bernauer, G. and Brocks, W. Micro-mechanical modelling of ductile damage and tearing-results of a European numerical round robin. *Fatigue and Fracture of Engineering Materials and Structures*, **25**: 363–384, 2002.

Casella, G. and Berger, R. *Statistical Inference*, 2nd edn. Duxbury Press, Belmont, CA, 2001.

Chakhunashvili, A., Barone, S., Johansson, P. and Bergman, B. Robust product development using variation mode and effect analysis. In Chakhunashvili, A., *Detecting, indentifying and managing sources of variation in production and product development*. PhD Thesis, Department of Quality Sciences, Chalmers University of Technology, Göteborg, 2006.

Davis, T. P. Science, engineering, and statistics. *Applied Stochastic Models in Business and Industry*, **22**: 401–430, 2006.

Davison, A. C. and Hinkley, D. V. *Bootstrap Methods and their Application*. Cambridge University Press, New York, 1997.

Ditlevsen, O. and Madsen, H. *Structural Reliability Methods*. John Wiley & Sons, Ltd, Chichester, 1996.

Efron, B. and Tibshirani, R. J. *An Introduction to the Bootstrap*. Chapman & Hall, New York, 1993.

Hjorth, U. *Computer Intensive Statistical Methods:Validation, Model Selection and Bootstrap*. Chapman & Hall, New York 1994.

Lodeby, K. *Variability analysis in engineering computational process*. Licentiate Thesis in Engineering, Mathematical Statistics, Chalmers University of Technology, Göteborg, 2000.

Melchers, R. *Structural Reliability Analysis and Prediction*, 2nd edn. John Wiley & Sons, Ltd, Chichester, 1999.

Pawitan, Y. *In All Likelihood: Statistical Modelling and Inference Using Likelihood*. Cambridge University Press, New York, 2001.

Pers, B.-E., Kuoppa, J. and Holm, D. Round robin test for finite element calculations. In A. F. Blom (editor), *Welded High Strength Steel Structures*, pp. 241–250. EMAS, London, 1997.

Svensson, T. Prediction uncertainties at variable amplitude fatigue. *International Journal of Fatigue*, **19**: S295–S302, 1997.

7

Predictive Safety Index for Variable Amplitude Fatigue Life

Thomas Svensson, Jacques de Maré and Pär Johannesson

7.1 Introduction

Reliability assessment with respect to fatigue failures is a difficult task, on the strength side because of the large scatter in fatigue life, on the load size due to highly varying product user profiles. An assessment procedure must therefore be simple enough to enable a quantification of vague input information, and be sophisticated enough to be a useful engineering tool for improvements.

Here, we present a trade-off between sophistication and simplicity that we think is suitable for applications sensitive to fatigue failures from rough structures such as welds or screw joints, primarily within the transport industry. The method is a development of the concept introduced in Karlsson et al. (2005), which was based on the classic load–strength method for fatigue at variable amplitude. The method uses scalar representatives for load and strength that are defined as equivalent amplitudes, and estimates their prediction uncertainties for reliability measures. The method may be seen as a first-order, second-moment reliability index according to the theory in Madsen et al. (1986), or a probabilistic Variation Mode and Effects Analysis (VMEA) according to Chapter 5.

One drawback that has been recognized in this method is its sensitivity to the choice of threshold in cycle counting of field data. Therefore we adapt here the idea of a fixed number of cycles related to an endurance limit. This will make the method more robust.

Another drawback in the method is that the slope in the Wöhler curve is assumed to be estimated for each application. In industrial practice, however, the slope is often determined by earlier experience and not from each fatigue test. Therefore, the uncertainty caused by the uncertainty in the slope is treated here explicitly.

In addition, the relationship to other reliability indices is outlined and the chosen level of approximation is discussed.

Robust Design Methodology for Reliability: Exploring the Effects of Variation and Uncertainty
edited by B. Bergman, J. de Maré, S. Lorén, T. Svensson
© 2009, John Wiley & Sons, Ltd

7.2 The Load–Strength Reliability Method

The reliability of a structure can be formulated as a comparison of the strength of the structure with the load that will act on it. For nonfailure we want to have the strength larger than the load during a certain target time or distance of usage,

$$\Sigma > \Lambda(t), \quad t < T_t,$$

where the strength Σ is modelled as a fixed value and the load Λ is modelled as a growing value that represents the damage accumulation. For the reliability calculation we will try to assess the condition at the target time T_t of usage by investigating if

$$\Sigma > \Lambda(T_t) \tag{7.1}$$

This will be done by defining a reliability index γ that is a function of the distribution of the strength and the final load measures, $\gamma = f(\Sigma, \Lambda(T_t))$. By putting some additional assumptions on the distribution of this reliability index, it is also possible to assess the probability of failure. However, the available information about the involved distributions is highly limited and assignments about failure probabilities, small enough to be interesting, often become misleading. Instead, we propose the direct use of the reliability index for comparisons and regard it as a measure of the distance to the failure mode.

This distance may be a combination of statistical measures, model uncertainties and engineering judgements, where the model uncertainties can be included in the statistical measure and the judgement parts can be estimated and updated based on experience or studies of similar constructions.

The main problem for reliability assessments in fatigue is to establish proper measures of the load and strength. These measures should be scalars in order to manage their statistical distributions in the reliability calculations. This implies a projection of the multidimensional load and strength features to one-dimensional spaces, which will introduce model errors. These must be taken into consideration by including them in the model uncertainty.

7.3 The Equivalent Load and Strength Variables

In fatigue, the service load can be seen as a random process of a time-varying load in each point of the structure to be analysed. The strength of the structure also depends on the time-varying characteristics of the load and we need a theory of damage accumulation to obtain the desired one-dimensional measures and to relate them to fatigue life. The established theory for the users is the Basquin equation combined with the Palmgren–Miner cumulative damage rule and we will here use this theory with certain practical modifications for the actual purpose of obtaining a useful reliability index.

The two main differences in the present method compared to the traditional use of damage accumulation theory are (i) material or component strength is assumed to be determined from variable amplitude tests in order to avoid large model errors caused by the cumulative damage rule, and (ii) a fictive endurance limit is defined which is suitably chosen to correspond to the required fatigue life of the structure. This limit reduces the sensitivity of the reliability measure to the omission level in the cycle counting technique.

We use a variant of the Basquin equation that is formulated with respect to an endurance limit for the component or material,

$$N = n_e \left(\frac{S}{\alpha_e}\right)^{-\beta}. \tag{7.2}$$

Here n_e is a suitably chosen cycle number corresponding to the endurance limit α_e and S is the load amplitude. A proper choice for steel constructions is $n_e = 10^6$. This formulation has the advantage that each property in the formula, except the exponent, has a physical interpretation; the parameter α_e has the same unit as the stress value and can be interpreted as an endurance limit, and the parameter n_e is a cycle number. The exponent β is assumed to be a material or component characteristic and is estimated by tests or by experience. The property α_e is regarded as a random variable that varies between specimens, reflecting different microstructural features and manifested by the observed scatter in fatigue tests. This random variable is defined as the *equivalent strength* and is modelled as a log-normal variable, where the parameters represent a certain component type or a specific material.

The strength of a component is estimated from laboratory test results. For spectrum tests the applied load consists of a variable amplitude time history that for damage analysis is described by its amplitude spectrum based on a cycle counting procedure (ASTM, 1999), giving the number of M applied load amplitudes, $\{S_i; i = 1, 2, \ldots, M\}$. According to the Palmgren–Miner cumulative damage assumption using Equation (7.2) we obtain:

$$D = \sum_{i=1}^{M} \frac{1}{N_i} = \sum_{i=1}^{M} \frac{S_i^{\beta}}{n_e \alpha_e^{\beta}} = \frac{1}{n_e \alpha_e^{\beta}} \sum_{i=1}^{M} S_i^{\beta},$$

where N_i is the life according to Equation (7.2) for the load amplitude S_i and M is the total number of counted cycles.

Failure occurs when D equals unity and the value of α_e for a specific specimen is:

$$\tilde{\alpha}_e = \left(\frac{1}{n_e} \sum_{i=1}^{M} S_i^{\beta}\right)^{1/\beta}, \tag{7.3}$$

which is an observation of the random equivalent strength variable. For the application of the load–strength model we need a comparable property for a service load spectra. This is constructed in a similar way to the calculated damage for the spectrum,

$$D_{T_d} = \sum \frac{1}{N_i} = \frac{T_d}{T} \sum_{i=1}^{M} \frac{L_i^{\beta}}{n_e \alpha_e^{\beta}} = \frac{\eta_L}{n_e \alpha_e^{\beta}} \sum_{i=1}^{M} L_i^{\beta}. \tag{7.4}$$

Here we sum over all counted amplitudes at L_i for the driven distance T, and T_d is the target life by means of, for instance, driving distance. For simplicity we also introduce the scaling factor $\eta_L = \frac{T_d}{T}$.

We now define the *equivalent load stress amplitude* L_{eq} as

$$L_{eq} = \left(\frac{\eta_L}{n_e} \sum_{i=1}^{M} L_i^{\beta}\right)^{1/\beta}, \tag{7.5}$$

and it can be seen from Equation (7.4) that

$$D_{T_d} = \frac{L_{eq}^{\beta}}{\alpha_e^{\beta}},$$

which equals unity when L_{eq} equals α_e. The distance between the logarithms of these two random properties represents a safety margin, which we will use for the reliability index. The equivalent load is modelled as a log-normal random variable.

7.4 Reliability Indices

The equivalent load and strength variables, L_{eq} and α_e, will now be used to form a second moment reliability index, i.e. a reliability index based only on expected values and variances of the scatter and uncertainty components.

One such index is the Cornell reliability index (see Madsen et al., 1986), which is based on the difference between the strength and the load, $\Sigma - \Lambda$,

$$\gamma_C = \frac{\mu_\Sigma - \mu_\Lambda}{\sqrt{\sigma_\Sigma^2 + \sigma_\Lambda^2}}, \tag{7.6}$$

where μ_Σ and μ_Λ are the expected values of the strength and load respectively, and σ_Σ^2 and σ_Λ^2 are their variances. Another similar index is based on a log transformation of the quotient Σ/Λ,

$$\gamma_{\log} = \frac{\mu_{\log \Sigma} - \mu_{\log \Lambda}}{\sqrt{\sigma_{\log \Sigma}^2 + \sigma_{\log \Lambda}^2}}, \tag{7.7}$$

In the literature about reliability indices a limiting function is often defined, $g(\cdot)$, which in the Cornell case is $g_1(\Sigma, \Lambda) = \Sigma - \Lambda$ and in the logarithmic case is $g_2(\Sigma, \Lambda) = \log \frac{\Sigma}{\Lambda}$.

The indices above are the normalized distances between the expected strength and the expected load in the linear and in the logarithmic scales, respectively. If the random variables are distributed according to the normal and the log-normal distributions, respectively, the indices determine the probability that the construction is safe.

In a more general situation the strength and the loads are vectors and not scalars and the border between the safe and the unsafe region, which in the cases above is a line that can very well be a curved surface. Then the normalized distances between the expectations do not determine the probability of safety even in the normal case.

One solution to this problem is the Hasofer–Lind reliability index (Madsen et al. 1986), which is based on a linearization of the limit function around a certain point at the limit surface, denoted as the *most probable point*, giving the smallest possible reliability index. However, this special point must normally be determined by a numerical minimization procedure, which requires knowledge about the g function for all scatter and uncertainty sources.

The present proposal is based on the Cornell reliability index in its logarithmic form (Equation 7.7), which is an adjustment to the observed linear behaviour of the fatigue phenomenon in logarithmic scale. By this transformation of the variables the nonlinearities are small with respect to the overall approximation level. The dispersion measure in the denominator will here be built from prediction uncertainties.

7.5 The Gauss Approximation Formula

In order to calculate proper prediction variances for the strength and load variables we will use the following approximation: For any function of random variables $f(x_1, x_2, \ldots, x_n)$, the variance of the function can be estimated as

$$Var[f(x_1, x_2, \ldots, x_n)] \approx \sum_{i=1}^{n} c_i^2 Var[x_i] + \sum_{i,j} c_i c_j Cov[x_i, x_j],$$

where the sensitivity coefficients are the partial derivatives,

$$c_i = \frac{\partial f}{\partial x_i}.$$

This approximation is based on the assumption that the function $f(\cdot)$ varies almost linearly with the variation of the influencing variables. In most cases the influencing variables are not correlated and then the formula is reduced to the first sum of weighted variances.

For the reliability index we will use the Gauss approximation formula repeatedly. For the strength variable there will, for instance, be variance contributions from at least material scatter, model errors, supplier variation and manufacturing tolerances. We can usually assume that these variables are uncorrelated and the uncertainty of the equivalent strength may be written,

$$
\begin{aligned}
\tau_{\ln \alpha_e} &= \sqrt{c_{test}^2 \cdot \sigma_{test}^2 + c_{model}^2 \cdot \sigma_{model}^2 + c_{suppl}^2 \cdot \sigma_{suppl}^2 + c_{man}^2 \cdot \sigma_{man}^2} \\
&= \sqrt{\tau_{test}^2 + \tau_{model}^2 + \tau_{suppl}^2 + \tau_{man}^2},
\end{aligned}
\tag{7.8}
$$

where we have included each sensitivity coefficient in the contributing uncertainty measure τ.

7.6 The Uncertainty Due to the Estimated Exponent β

A crucial parameter in the load–strength interaction in fatigue is the Wöhler exponent. This parameter is in industrial practise often fixed at a specific value, for instance $\beta = 3$ for welds, $\beta = 5$ for vaguely defined structures or $\beta = 8$ for high-strength components with smooth surfaces. Experience and historical data may contain information about the uncertainty in such fixed values, the weld exponent may for instance be regarded to be uncertain within the limits $\{2.5; 3.5\}$. In other cases the exponent is estimated from specific tests and the uncertainty can be determined directly from a statistical analysis of the test result. We will now investigate how much the uncertainty in β influences the uncertainty in the distance between load and strength.

For brevity, we introduce the following spectrum notation:

$$E[S^\beta] = \frac{1}{M} \sum_{i=1}^{M} S_i^\beta,$$

where we sum over all load amplitudes S_i in the spectrum until failure occurs. Then we can write for jth sample equivalent strength,

$$\tilde{\alpha}_{e,j}^\beta = \frac{1}{n_e} \sum_i S_{j,i}^\beta = \frac{N_{S,j}}{n_e} E_j[S^\beta] \tag{7.9}$$

where the test number j was performed with the spectrum $E_j[S^\beta]$ until failure occurred.

Further, we assume that a number of load measurements in service have been performed and their equivalent loads are written:

$$L_{eq,j}^{\beta} = \frac{\eta_{L,j} N_{L,j}}{n_e} E_j[L^{\beta}] = \frac{n_{T,j}}{n_e} E_j[L^{\beta}],$$

where the load measurement number j contained $N_{L,j}$ cycles of the spectrum $E_j[L^{\beta}]$, and we introduce the target life in cycles: $n_T = \eta_{L,j} N_{L,j}$ for all j. For the estimated reliability index distance we obtain,

$$\overline{\ln \alpha_e} - \overline{\ln L_{eq}} = \frac{\langle \ln E[S^{\beta}] \rangle - \ln n_e + \overline{\ln N_S} - (\langle \ln E[L^{\beta}] \rangle - \ln n_e + \ln n_T)}{\beta}$$

$$= \frac{\langle \ln E[S^{\beta}] \rangle - \langle \ln E[L^{\beta}] \rangle + \overline{\ln N_S} - \ln n_T}{\beta},$$

(7.10)

where averages are denoted by bars or the brackets $\langle \cdot \rangle$.

In order to find the uncertainty contribution to the reliability distance from the parameter β, we use the Gauss approximation formula and calculate the sensitivity coefficient by differentiating the result (Equation 7.10) with respect to β:

$$\frac{\partial(\overline{\ln \alpha_e} - \overline{\ln L_{eq}})}{\partial \beta} = -\frac{\langle \ln E[S^{\beta}] \rangle - \langle \ln E[L^{\beta}] \rangle + \overline{\ln N_S} - \ln n_T}{\beta^2}$$

$$+ \frac{1}{\beta} \left(\left\langle \frac{E[S^{\beta} \ln S]}{E[S^{\beta}]} \right\rangle - \left\langle \frac{E[L^{\beta} \ln L]}{E[L^{\beta}]} \right\rangle \right)$$

$$= \frac{1}{\beta} \left(\frac{1}{\beta}(\ln n_T - \overline{\ln N_S}) + \left\langle \frac{E[S^{\beta} \ln S]}{E[S^{\beta}]} \right\rangle - \frac{1}{\beta} \langle \ln E[S^{\beta}] \rangle \right.$$

(7.11)

$$\left. - \left\langle \frac{E[L^{\beta} \ln L]}{E[L^{\beta}]} \right\rangle + \frac{1}{\beta} \langle \ln E[L^{\beta}] \rangle \right)$$

$$= \frac{1}{\beta} \left(\frac{1}{\beta} \ln \frac{n_T}{\tilde{N}} + \bar{\xi}_S - \bar{\xi}_L \right)$$

where $\tilde{N}$ is the geometric average of the fatigue lives of the reference specimens, and $\bar{\xi}_S$ and $\bar{\xi}_L$ are the averages of numbers representing spectrum type.

The exponent sensitivity depends on the difference between the spectrum type numbers for the reference test and for the usage, respectively.

The shape measure for a certain spectrum, $\xi_X = \frac{E[X^{\beta} \ln X]}{E[X^{\beta}]} - \frac{1}{\beta} \ln E[X^{\beta}]$ is scale invariant and is equal to zero for a constant amplitude load, see Appendix.

7.7 The Uncertainty Measure of Strength

In Equation (7.3) we defined the equivalent strength for a certain specimen,

$$\tilde{\alpha}_e = \left(\frac{1}{n_e} \sum_{i=1}^{N} S_i^{\beta} \right)^{1/\beta},$$

where the random property is the accumulated damage to failure, while the spectrum is supposed to be known and the value n_e is a fixed constant. Assuming that the life is log-normally distributed, we calculate the standard deviation of the logarithmic observed equivalent strengths,

$$m_S = \frac{1}{n} \sum_{i=1}^{n} \ln \tilde{\alpha}_{e,i}, \qquad s_\alpha = \sqrt{\frac{1}{n-1} \sum_{i=1}^{n} (\ln \tilde{\alpha}_{e,i} - m_S)^2},$$

where n is the number of fatigue tests.

From this estimated standard deviation we want to obtain a measure of prediction uncertainty for the reliability index. We will do that by using a proper 95% prediction interval for a future assessment of the strength, given the test. If we assume that the logarithm of the equivalent strength is normally distributed, such a prediction interval can be calculated from the Student-t distribution,

$$m_S \pm s_\alpha \cdot t_{0.025,n-1} \cdot \sqrt{1 + \frac{1}{n}}.$$

Here, the t-value compensates for the uncertainty in the standard deviation estimate and the one over n term under the root sign takes the uncertainty in the mean value m_S into account.

Since the t-value approaches approximately 2 when the number of tests increases to infinity, we finally define the actual uncertainty component for the reliability index,

$$\delta_{S,1} = \frac{s_\alpha \cdot t_{0.025,n-1}}{2} \cdot \sqrt{1 + \frac{1}{n}},$$

which thereby is a measure corresponding to a standard deviation of the predicted equivalent strength.

This initial uncertainty measure includes the scatter of the specimens, the uncertainty in the estimated mean and the uncertainty in the variance estimate. The uncertainty due to the exponent β, treated in the previous section, influences both strength and load and will be added separately. The influence of the exponent uncertainty on the calculated standard deviation is neglected.

However, when comparing the strength with the load in future there may be additional variation and model error sources involved. For instance, one must take into account that the tested specimens are taken from a single supplier, while several suppliers may be used in future. Such a source of variation is difficult to quantify, but by experience one might be able to assess its influence by, say, 10%. Since we work here with natural logarithms such a judgement can be quantified directly as a standard deviation contribution,

$$\delta_{S,2} = 0.10.$$

Apart from variation sources such as suppliers there are also other uncertainty sources. One is due to the lack of equivalence between laboratory reference tests and the service situation. Multidimensional loads, environment, multiaxiality, residual stresses or mean values are different in laboratory tests compared with service conditions and such model errors should be represented in the reliability index. By experience, specially designed tests and judgements these sources may also be quantified by one or more percentage variation, resulting in estimates, $\delta_{S,i}$, where the index $\{i = 3, \ldots, m\}$ indicates different identified sources. By putting more effort on test design and use components or structures instead of material specimens, these uncertainties can be reduced and balanced to the cost of the additional work.

The damage accumulation theory including the Basquin equation and the formulation of the equivalent load is an approximation and will contribute to the uncertainty in life prediction. Just like the problem above, contributions such as this may be estimated by experience, literature data, specially designed tests and so forth, adding more percentage variation components $\delta_{S,i}$ $\{i = m + 1, \ldots, p\}$. By performing a laboratory test using load spectra with similar properties as the ones in service, these model uncertainties can be reduced at the cost of more advanced tests.

In total, the uncertainty measure of strength is calculated by a simple addition of the corresponding variances of log strength,

$$\delta_S^2 = \sum_{i=1}^{p} \delta_{S,i}^2.$$

The quadratic summation results in an emphasis of the largest contributions and in practice all components that are less than 20% of the largest component can be neglected.

7.8 The Uncertainty Measure of Load

The load variation originates from the population of users. This may be identified as customers, markets, missions or owners depending on the application. In the case of personal cars it may be necessary to consider the whole population of users, while for trucks it is desirable to distinguish between populations with common missions and environments, such as timber transport, city distribution or highway usage. For a front loader or an excavator the population may be divided into, for instance, road construction and mining. For each population, different reliability indices can be calculated. The important thing is that it is made clear what population the reliability index relates to.

Once the population has been defined, the mean and variance of the corresponding load need to be estimated. This is often quite a difficult task. In the automotive industry there is often an established relation between the assumed population and the company specific test track. In military air space applications special predefined missions are established, and in other industrial practises target loads are specified by experience and sometimes given as standardized load spectra. These load specifications are rather rough approximations of reality and usually only represent the mean user or a certain extreme user.

Because of the extremely fast development of information technology, many companies now work with direct measurements of loads in service, which makes it possible to obtain the properties needed for a proper reliability assessment. However, this usually demands large measurement campaigns, where the sampling of customers, environments and markets is a subtle matter, demanding great care when planning the campaigns (see Karlsson, 2007). From such measurements on a specific population it is possible to calculate equivalent loads according to Equation (7.5) for each sample and find the mean and the standard deviation of the logarithmic transformation of the population,

$$m_L = \frac{1}{n} \sum_{i=1}^{n} \ln L_{eq,i}, \qquad s_L = \sqrt{\frac{1}{n-1} \sum_{i=1}^{n} (\ln L_{eq,i} - m_L)^2}.$$

If the number of samples is small, there will be an uncertainty in the estimated standard deviation and our uncertainty measure must be adjusted as in the strength case, again assuming a normal distribution for our logarithmic property,

$$\delta_{L,1} = s_L \frac{t_{0.975,n-1}}{2} \sqrt{1 + \frac{1}{n}}.$$

If the load measurements have been made on a test track or if the measurement campaign is known not to be a random sample of the actual population, there may be an unknown bias in the estimates and by judgement more uncertainty components should be introduced,

$$\delta_L = \sqrt{\delta_{L,1}^2 + \delta_{L,2}^2 + \delta_{L,3}^2 + \cdots},$$

where each different component $\delta_{L,i}$ is estimated as the coefficient of variation of equivalent load, i.e. based on a percentage judgement.

7.9 The Predictive Safety Index

On the basis of the logarithmic variant of the Cornell reliability index (Equation 7.7) and the properties derived above we now formulate the predictive safety index for fatigue,

$$\gamma_p = \frac{m_S - m_L}{\sqrt{\delta_S^2 + \delta_L^2 + \delta_\beta^2}}.$$

As this index is based on the logarithm of the strength and loads it can be interpreted as a weighted safety factor and be used in the same manner as a traditional safety factor, not being related to a measure of probability. In that case the target value of the index must be decided by experience; initially it may be based on established safety factors and subsequently updated with new information from failures. The advantage of this index compared to a deterministic safety factor is that it is clearly related to the scatter and uncertainty sources involved. The influence of the index from, for instance, more measurements in service, model improvements by experiments, or more specified component tolerances can easily be monitored and judged against costs and other efforts.

By placing distributional model assumptions on the reliability index, it is possible to calculate theoretical probabilities of failure, and requirements about, for instance, one failure in ten thousand cases can be judged. In fact, if the mean values are normally distributed, then the reliability index may be interpreted as a percentile in the normal distribution and it is easy to calculate the corresponding probability of failure. However, in order to interpret such probabilities as true failure frequencies it is essential that the tails of the distributions follow the assumed model and this is normally not the case for small probabilities. Therefore, such theoretical probabilities should be used only as comparisons between different solutions and possibly as a starting point for deciding a target value for the index.

7.10 Discussion

The use of 95% prediction limits for obtaining the index reflects the ideas above. In this part of the distribution it is often possible to adopt the normal distribution assumption, and the procedure using the *t*-distribution corrections is motivated. The normal distribution assumption at this stage corresponds to the assumption of a log-normal distribution on fatigue life, which in turn can be motivated by the observed linear behaviour of the Wöhler diagram.

Other load–strength procedures for fatigue recommend the use of the three-parameter Weibull distribution for the strength and the reversed three-parameter Weibull for the load. This suggestion is motivated by the fact that the fatigue strength can be regarded as a weakest link property. In addition, this approach takes into account the possibility of a lower limit for strength and an upper limit for load, which is appealing. There are two arguments against this Weibull approach: (i) it is difficult to include the statistical uncertainty in the estimated Weibull parameters, which means that the gains with more tests are hidden; (ii) the estimate of the threshold parameter in the three-parameter Weibull is highly uncertain in the case of the small sample sizes that are used in fatigue.

Furthermore, the difference using a three-parameter Weibull and a log-normal distribution vanishes when uncertainties originating from model errors, material suppliers and lack of test equivalence are added. The use of a threshold is possible also in the log-normal case, but in any case such a threshold should be estimated from physical considerations, and not by statistical estimates.

Appendix

The spectrum type measure is

$$\xi_S = \frac{E[\ln S \cdot S^\beta]}{E[S^\beta]} - \frac{\ln E[S^\beta]}{\beta}.$$

We evaluate the measure for a constant amplitude load:

$$\xi_{S(CA)} = \frac{E[\ln S \cdot S^\beta]}{E[S^\beta]} - \frac{\ln E(S^\beta)}{\beta} = \frac{\ln S \cdot S^\beta}{S^\beta} - \frac{\ln S^\beta}{\beta} = \ln S - \ln S = 0.$$

Further, we investigate the influence of a scaling factor on a variable amplitude spectrum:

$$\begin{aligned}
\xi_{cS} &= \frac{E[\ln cS \cdot (cS^\beta)]}{E[(cS)^\beta]} - \frac{\ln E[(cS)^\beta]}{\beta} \\
&= \frac{c^\beta E[\ln c \cdot S^\beta] + c^\beta E[\ln S \cdot S^\beta]}{c^\beta E[S^\beta]} - \frac{\ln c^\beta + \ln E[S^\beta]}{\beta} \\
&= \frac{\ln c E[S^\beta] + E[\ln S \cdot S^\beta]}{E[S^\beta]} - \ln c - \frac{\ln E[S^\beta]}{\beta} \\
&= \frac{E[\ln S \cdot S^\beta]}{E[S^\beta]} - \frac{\ln E[S^\beta]}{\beta} \\
&= \xi_S.
\end{aligned}$$

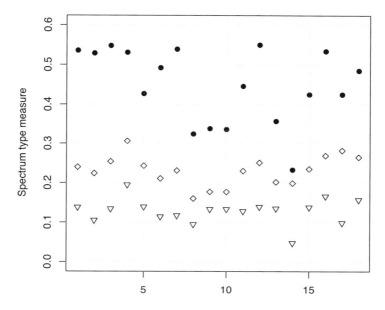

Figure 7.1 The spectrum type measure for eighteen different road measurements. The lowest values (inverted triangles) represent a 30% omission level and the highest 10% (filled circles).

As can be seen the shape measure is scale invariant and equal to zero for a constant amplitude load.

Numerical investigations of the type measure ξ show that it is quite sensitive to the omission level in the counted spectra. The omission level in practical use is usually between 10% and 30% of the maximum load amplitude as amplitudes far below the fatigue limit are assumed to be nondamaging. In Figure 7.1, the spectrum type measure is plotted for a number of different road measurement spectra using three different omission levels, 10%, 20% and 30%. For Gaussian and linear spectra with a 10% omission level, one obtains the measures 0.22 and 0.29 respectively. Here, the terms 'Gaussian' and 'linear' mean that the amplitudes counted are given by the Weibull distribution,

$$P(S > X) = \exp\left[-\left(\frac{x}{a}\right)^b\right].$$

where the shape parameter b equals two for the Gaussian spectrum and equals one for the linear spectrum. The example shows that the omission level highly influences the spectrum type measure, giving a lower number.

Example

If, for instance, the reference tests have a geometric mean life five times higher than the target life and the exponent β is equal to 3, then the first term is

$$\frac{1}{\beta}\ln\frac{n_T}{\tilde{N}} = \frac{\ln 5}{3} = 0.54$$

If the difference between the spectrum type measures is, say -0.4, we obtain the sensitivity coefficient:

$$\frac{1}{\beta}\left(\frac{1}{\beta}\ln\frac{n_T}{\tilde{N}}+\xi_S-\xi_L\right)=\frac{0.54-0.4}{3}=0.05.$$

To obtain the uncertainty component regarding the exponent, we should multiply this value by the uncertainty of the exponent itself. Assume that the exponent is judged to be contained in the interval $[2.5; 3.5]$. Assume further that the uncertainty distribution for the exponent is uniform. Then the variance is

$$\frac{(3.5-2.5)^2}{12}=0.083$$

The uncertainty measure for the reliability index is

$$\delta_\beta=\sqrt{0.05^2\cdot0.083}=0.064,$$

i.e. the uncertainty due to the exponent is about 6%.

References

ASTM. Standard practises for cycle counting in fatigue analysis, *ASTM E 1049-85. Annual Book of ASTM Standards*, 0301: 710–718. American Society for Testing and Materials, West Conshohocken, PA, 1999.

Karlsson M. *Load modelling for fatigue assessment of vehicles—a statistical approach*, PhD thesis, Mathematical Statistics, Chalmers University of Technology, 2007.

Karlsson, M., Johannesson, B., Svensson, T. and de Maré, J. Verification of safety critical components. Presentation at the *VDI Conference: Trucks and Buses - Solutions of Reliability, Sustainable Environment and Transport Efficiency*. Böblingen, Germany, June 2005.

Madsen, H.O., Krenk, S. and Lind, N.C. *Methods of Structural Safety*. Prentice-Hall, NJ, 1986.

8

Monte Carlo Simulations Versus Sensitivity Analysis

Sara Lorén, Pär Johannesson and Jacques de Maré

8.1 Introduction

Monte Carlo simulations have become very popular in industry where the influence of variation is examined. The aim of this chapter is to compare the crude Monte Carlo simulation concept with the method of linearization using only sensitivity coefficients. This is a method of moments (see Morrison, 1998; Johansson et al., 2006.), which is the main tool in Chapter 6 and 7. The linearization method is in some contexts called the first-order second-moment method. The term 'first order' relates to the fact that a linearization of the function is used, and the term 'second moment' denotes that it is based only on expected values and variances.

The transfer function is the function from the input vector x to the output y, $y = f(x)$. In industrial applications the transfer function is often given implicitly, e.g. the evaluation may include a finite element (FE) analysis. Therefore, each evaluation of the transfer function may take several minutes, or even hours to perform. In these cases it is not feasible to use the full transfer function when the output is calculated. The solution is often to calculate an approximation to the transfer function, for example using splines, quadratic functions, linearization, or some other explicit expressions to approximate the true transfer function.

There are many differences between the methods and an important one is what information about the input variables, x, is needed. For the Monte Carlo simulation the complete statistical distribution of x is needed and has to be known or very often guessed in some intuitive way. This distribution will in some cases have a very significant influence on the result, which will be shown later. For the linearization method, only the mean and the standard deviation of the input variables is used.

For the Monte Carlo simulation, the input variables are simulated from the chosen input distribution, and by using the approximation of the transfer function, $f(x)$, the output y is

Robust Design Methodology for Reliability: Exploring the Effects of Variation and Uncertainty
edited by B. Bergman, J. de Maré, S. Lorén, T. Svensson

calculated. This is repeated a large number of times and an approximate distribution of y is estimated.

For the linearization method, a linearization about the mean is performed and the sensitivity coefficients are used. This can be done in different ways, taking the derivatives of the approximation $f(x)$, using the estimated regression coefficients of $f(x)$ or using the differential quotients from two different evaluations – in this case the transfer function does not need to be estimated. After that a weighted summation of the squares of the sensitivity coefficients leads to a variance for y, i.e. the complete statistical distribution for y is not estimated.

The two methods will be evaluated in an example from an industrial context.

8.2 Transfer Function

The transfer function is needed for the Monte Carlo simulation and can be used for the linearization method as mentioned above. The transfer function has to be estimated in some way, e.g. by design of experiment. An approximate transfer function that consists of both interaction terms and second-order terms is as follows:

$$y = \beta_0 + \sum_{i=1}^{I} \beta_i \cdot x_i + \sum_{i=1}^{I-1} \sum_{j=i+1}^{I} \beta_{ij} \cdot x_i \cdot x_j + \sum_{i=1}^{I} \beta_{ii} \cdot x_i^2, \tag{8.1}$$

where I is the number of input variables. This transfer function will be denoted as $y = f_2(x)$, and if only the first-order terms are used, as $y = f_1(x)$. The different coefficients, $\hat{\beta}_i$ and $\hat{\beta}_{ij}$, are estimated using linear regression.

For the Monte Carlo simulation, each x_i is simulated according to its statistical distribution and then y is calculated using the estimated transfer function. This is repeated a number of times and a distribution function for y is estimated.

For the linearization method the uncertainty of the output variable y is estimated and not the complete statistical distribution. The Gauss approximation formula is used in the following way:

$$Var(y) \approx \sum_{i=1}^{I} c_i^2 Var(x_i) + \sum_{i \neq j} c_i c_j \cdot Cov(x_i, x_j), \tag{8.2}$$

where c_i is the sensitivity coefficient belonging to x_i. Under the assumption that the input variables are independent, the variance of y is simplified to

$$Var(y) \approx \sum_{i=1}^{I} c_i^2 Var(x_i). \tag{8.3}$$

The sensitivity coefficients $c_1, c_2, \cdots, c_I$ are the partial derivatives of the transfer function with respect to the input variables, $c_i = \partial f / \partial x_i$. The linearization is around the mean of the input variables, $\mu_i = E[X_i]$. The sensitivity coefficient for x_i using the transfer function in Equation (8.1) is

$$c_i = \hat{\beta}_i + \sum_{i \neq j} \hat{\beta}_{ij} \cdot \mu_j + 2\hat{\beta}_{ii} \cdot \mu_i. \tag{8.4}$$

For a linear transfer function the sensitivity coefficients is just the estimated regression coefficients, $c_i = \hat{\beta}_i$. How good the approximations in Equations (8.2) and (8.3) are depends on the properties of the transfer function. Is it linear, roughly linear or highly nonlinear?

For a non-linear transfer function f_2, the mean of the output variable y is $E[Y] \approx f_2(\mu)$. Again the approximation depends on the properties of the transfer function. For the linear transfer function the mean is $E[Y] = f_1(\mu)$.

The work that remains after the transfer function is estimated is different for the two methods. For the Monte Carlo simulation the complete simulation is left and for the linearization method only a weighted summation is to be calculated.

8.3 Example from Industrial Context

The component that is analysed is taken from Karlén (2007) and is a part of a nozzle and is shown in Figure 8.1. In Karlén (2007) six nodes were chosen for analysing the fatigue life. The most critical node was identified and that node will be analysed here. In Figure 8.1 the critical point is indicated.

The four different categories of uncertainty sources that will be used to describe the life are the geometry with six parameters (numbers 1 to 6 in Figure 8.1), temperatures with three parameters for the low cycle fatigue (LCF) curve, four different physical properties and four different thermal properties. In the original work these were to be modelled and evaluated separately, because of the limited computational time. For a variety of reasons the geometry parameters 7 to 12 were not analysed (see Karlén, 2007).

The output variable considered is the life, which is given as a function of the input, $N = f(x)$. The unit for the life will be main life cycle (MLC). One MLC consists of two ground tests, one aborted flight and a flight. In Karlén (2007) it was found that this transfer function was nonlinear and contained interaction terms, which resulted in a skewed distribution for the life, i.e. it could not be normally distributed but some of the assumptions are built on normality.

However, the log-life is often better modelled by a normal distribution, and hopefully contains less nonlinearities. Therefore the log-life transfer function is used here, $\ln(N) = g(\ln(x))$. This is illustrated for the thermal source in Figures 8.2 and 8.3. The calculated lives

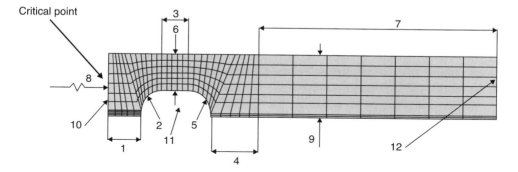

Figure 8.1   Geometry of the analysed component (see Karlén, 2007).

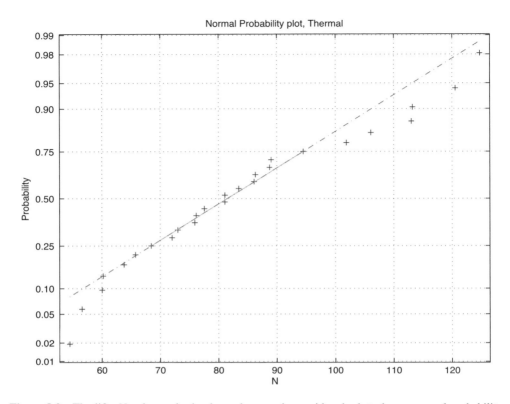

Figure 8.2 The life, N, when only the thermal source is considered, plotted on a normal probability plot.

are plotted on a normal probability plot with and without using the logarithm. From the figures it looks as if the logarithmic life follows a normal distribution more than the life itself.

8.3.1 Result from Karlén (2007)

In Karlén (2007) the mean and the standard deviation for the life in MLC were estimated for each source using Monte Carlo simulations. This is shown in Table 8.1. The logarithm of the life will be used here instead of the life itself, as mentioned before. Another advantage of using the logarithm is that its variance can be interpreted in terms of the coefficient of variation, W, for the life, i.e.

$$Var(\ln(N)) \approx \frac{Var(N)}{(\mathrm{E}[N])^2} = W^2. \tag{8.5}$$

From the Monte Carlo simulations in Karlén (2007) the coefficient of variation, W, for each source is calculated using Equation (8.5) and the squared W is assumed to be the variance for the logarithmic life. This is shown in Table 8.2.

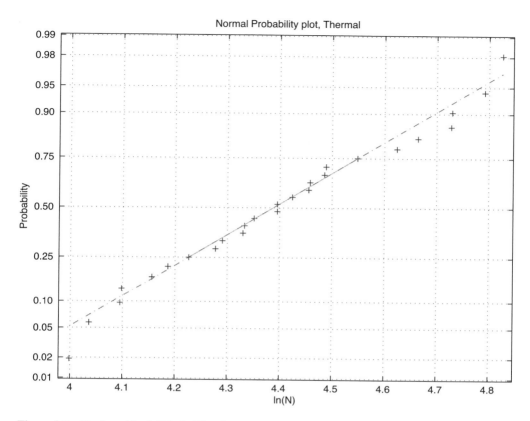

Figure 8.3 The logarithmic life, $ln(N)$, when only the thermal source is considered, plotted on a normal probability plot.

8.3.2 Different Transfer Functions and Input Distributions

Each of the four sources has a number of variables x_i and nominal values x_{nom_i} which are used to produce normalized entities, $z_i = (x_i/x_{nom_i})$. In Karlén (2007) three levels for each variable z_i were used to calculate the life. The Box–Behnken design was used and is based on a two-level factorial design with an incomplete block design and then a specified number of replicated centre points are added.

Table 8.1   Result from Monte Carlo simulations in Karlén (2007).

Source	$std(N)$	E [N]
Geometry	7.70	81.93
LCF	4.81	81.86
Physical	2.22	81.11
Thermal	8.53	81.61

Table 8.2 Sources of the $Var(\ln(N))$ using Equation (8.5) and result in Karlén (2007).

Source	Geometry	LCF	Physical	Thermal
$\hat{Var}(\ln(N))$	0.0088	0.0035	0.0007	0.0109

From the Finite Element Model (FEM) calculation in Karlén (2007) two different approximative transfer functions were estimated that use the logarithm both on the normalized input parameters and the output parameter. The first one uses only the first-order terms and is $\ln(N) = g_1(\ln(z))$. The second also uses interaction and second-order terms and is as follows:

$$\ln(N) = \hat{\beta}_0 + \sum_{i=1}^{I} \hat{\beta}_i \cdot \ln(z_i) + \sum_{i=1}^{I-1} \sum_{j=i+1}^{I} \hat{\beta}_{ij} \cdot \ln(z_i) \cdot \ln(z_j) + \sum_{i=1}^{I} \hat{\beta}_{ii} \cdot \ln(z_i)^2 \quad (8.6)$$

where I is the number of variables of the source. The function in Equation (8.6) will be denoted $\ln(N) = g_2(\ln(z))$.

The variation of the input variables for the geometry source was varied according to Table 8.3 and is taken from Karlén (2007). The variation is assumed to be at the 3σ level and corresponds approximatively to a failure risk of 1 in 1000, often the risk of interest for this type of application.

For the Monte Carlo simulations, the estimated life distribution depends both on the chosen approximative transfer function and on the chosen input distribution. To illustrate this, the input variables are simulated from three different distributions, normal, log-normal and Weibull all with the same mean, $E[Z] = 1$, and variances according to Table 8.3. Using the two transfer functions g_1 and g_2 combined with the three input distributions, it is possible to estimate six different life distributions using the Monte Carlo simulation. In Figure 8.4 the six different estimated life distributions from 100,000 simulations are shown.

As can be seen in Figure 8.4 the shape of the life distribution depends very much on the chosen input distribution. The largest difference is in the tail of the distributions and this is the most interesting part for this application and for many other applications as well. In Table 8.4 some of the percentiles are shown and the main difference between them is for the low and high percentiles.

The estimated standard deviation for the logarithmic life is approximately the same for all six different estimated life distributions, 0.09. But the estimated mean of the calculated logarithmic

Table 8.3 Variation for each input parameter for the geometry source.

Parameter	Variation (%)
Thickness of the hot wall, x_1	±25
Inner radius, x_2	±37.5
Height of the mid-wall, x_3	±15
Outer thickness, x_4	±16.67
Outer radius, x_5	±10
Thickness of the mid-wall, x_6	±10

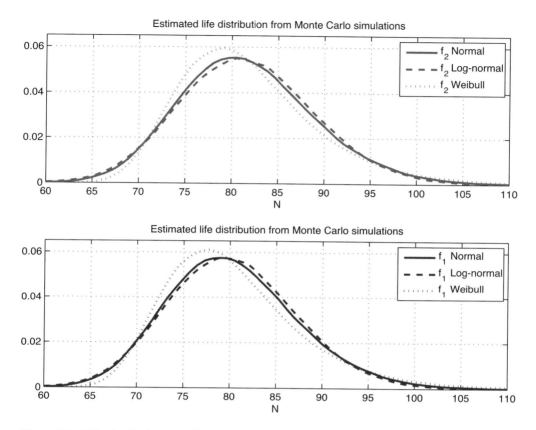

Figure 8.4   Life distribution according to the geometry source for the two transfer functions g_1 and g_2 with three different input distributions.

life is not exactly the same for all the cases considered. For the three life distributions where g_2 (second-order model) is used as a transfer function it is 4.40. For the other transfer function, g_1 (first-order model), the mean is 4.38 for the logarithmic life.

This result indicates in some sense the problem with the Monte Carlo simulations. It is easy to believe too much in the tail of the output distribution since it looks very correct and

Table 8.4   Different life percentiles utilizing a transfer function that uses the logarithm on both the input and output variables, g_i.

Percentile	Normal		Log-normal		Weibull	
	g_2	g_1	g_2	g_1	g_2	g_1
0.05th	62	62	60	60	65	65
1th	67	66	66	66	69	68
50th	81	80	81	80	81	79
99th	102	100	100	98	105	104

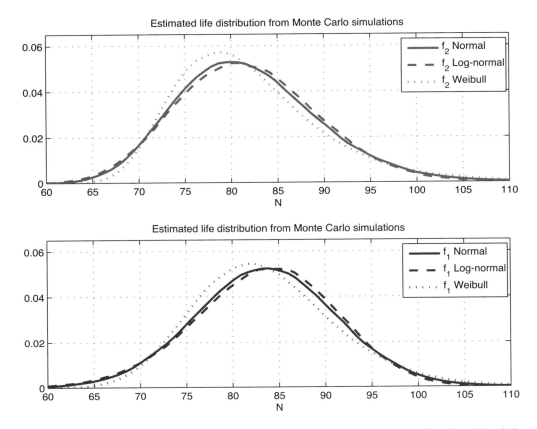

Figure 8.5   Life distribution according to the geometry source for the two transfer functions f_1 and f_2 with three different input distributions.

precise, but it depends very much on the input distribution and the transfer function. One question that needs to be asked is whether the chosen input distribution really is the correct distribution.

In Karlén (2007), transfer functions without the logarithm were used. Here also these types of transfer functions will be used, f_1 (first-order model) and f_2 (second-order model). Again the reason is to see how the choice of transfer function and input distribution affects the result. The same three input distributions as before are used and another six life distributions are estimated. These distributions are shown in Figure 8.5 and the percentiles in Table 8.5. It is clear that in these simulations also the chosen transfer function affects the life distribution. The estimated mean is 83.5 and the standard deviation of the life is 7.6 if only first-order terms are used, f_1. For the other transfer function f_2 (second-order model) the mean is 81.8 except if a normal distribution is used as an input distribution; then it is 81.9 and the standard deviation of the life is between 7.5 and 7.9 depending on the chosen input distribution.

Table 8.5 Different life percentiles using a transfer function without the logarithm on the input and output variables, f_i.

	Normal		Lognormal		Weibull	
Percentile	f_2	f_1	f_2	f_1	f_2	f_1
0.05th	63	58	61	56	65	64
1th	67	66	66	65	69	69
50th	81	84	81	84	81	83
99th	102	101	101	100	106	104

8.3.3 Linearization Method Using Different Transfer Functions

For the linearization method only the variance of the input variables is used. Therefore it is not necessary to know (or estimate) the statistical input distributions. Only the variance for the life is estimated and not the full distribution as for the Monte Carlo simulation. This is done by calculating the sensitivity coefficient for each input variable, c_i, from the estimated transfer function and combining it with the variances from the input variables. The total variation for the life from one source is calculated using a weighted summation in the following way

$$Var(N) = \sum_{i=1}^{I} \tau_i^2 = \sum_{i=1}^{I} c_i^2 \cdot Var(x_i).$$ (8.7)

In Table 8.6 the different τ_i are shown for the different transfer functions together with the standard deviation for the life or the standard deviation for the logarithmic life depending on the chosen transfer function. In this example the mean is the same for the normalized entities z_i. The three levels for the input variables used in the linear regression are chosen symmetrically around the mean, which here is 1. This is why the first-order model f_1 and the second-order model f_2 give exactly the same result. The results for the two functions g_1 and g_2 are approximately the same.

From the different estimated τ_i it is clear that for the geometry source that it is the parameters x_1 and x_2 that contribute the most to the variation in life. This is one advantage of the linearization method; it is easy to identify which input variable in the source contributes most

Table 8.6 The τ_i for all geometry parameters using the four transfer functions described.

Parameter	f_1, f_2	g_1	g_2
x_1	5.39	0.062	0.062
x_2	5.17	0.058	0.061
x_3	0.88	0.011	0.011
x_4	0.05	0.0007	0.0006
x_5	0.30	0.003	0.003
x_6	0.94	0.012	0.012
Standard deviation	$std(N) = 7.6$	$std(\ln(N)) = 0.09$	$std(\ln(N)) = 0.09$

Table 8.7 Different life percentiles using $\widehat{std}(N) = 7.6$.

	Percentile			
Estimated mean of N	0.05th	1th	50th	99th
$\hat{N}_{FEM} = 81.3$	56.3	63.6	81.3	99.0
$\hat{N}_{f_1} = 83.5$	58.5	65.8	83.5	101.2

to the variation of the output and thereby know where to put the effort to decrease the variation (if it is possible).

After the transfer function is estimated, much of the work with estimating the variance for the output variable has been completed for the linearization method. The method is built on linearization around the mean of the input variables. Therefore if the model is nonlinear it is worth some effort to see if there is some transformation of the data that makes the model linear, at least linear in the part of the function of interest. A useful transformation to obtain linearity is the logarithmic function.

8.3.4 Comparing the Results from the Two Methods

In Tables 8.4 and 8.5 the percentiles for the life using Monte Carlo simulations were shown. To compare the two methods the corresponding percentiles will be estimated using the other method, the linearization method. This is done by assuming normality for the life or the logarithmic life. The estimated standard deviation for the life and the logarithmic life is taken from Table 8.6, i.e. $std(N) = 7.6$ and $std(\ln(N)) = 0.09$.

Two different estimated means for the life will be used. The first one will be denoted $\hat{N}_{FEM}$ and is the calculated life from the FEM using the nominal input variables, $\hat{N}_{FEM} = 83.5$. The calculated life using the estimated transfer function f_1 with nominal input variables is the other mean and is $\hat{N}_{f_1} = 81.3$. If the logarithmic life is used in the analysis the corresponding means are $\ln(\hat{N}_{FEM}) = 4.40$ and $\ln(\hat{N}_{g_1}) = 4.38$, respectively.

In Table 8.7 the life percentiles are estimated for different percentiles with the two different estimated means but with the same standard deviation, $\widehat{std}(N) = 7.6$. For the logarithmic analysis the percentiles for the logarithmic life is transformed to life and are shown in Table 8.8. Again two different means are used and the standard deviation for the logarithmic life is, $\widehat{std}(\ln(N)) = 0.09$.

The linearization method is built on the normal assumption and some of the estimated life distributions from the Monte Carlo simulations have a normal shape, see Figures 8.4 and 8.5. In these cases the Monte Carlo simulations and the linearization method give a similar result, see Tables 8.4, 8.5, 8.7 and 8.8.

In the case where the transfer function is linear and the input distribution is normal, the two methods will give the same answer. This is the case when f_1 is used together with a normal input distribution. If log-normal distributed input variables are used on the other linear function g_1, the logarithmic life is normally distributed and the two methods should give the same result. The reason why the result differs between the two methods here is that $E[X] = 1$ and not $E[\ln(X)] = 0$.

Table 8.8 Different life percentiles using $\widehat{std}(\ln(N)) = 0.09$.

Estimated mean of $\ln(N)$	Percentile			
	0.05th	1th	50th	99th
$\ln(\hat{N}_{FEM}) = 4.40$	60.6	66.1	81.5	100.4
$\ln(\hat{N}_{g_1}) = 4.38$	59.4	64.8	79.8	98.4

For the nonlinear transfer function, the difference between the two methods depends on how linear the function is. Is the function roughly linear or highly nonlinear? This is because the remainder term in the Gauss approximation is fairly small for a roughly linear function and can be fairly large for a highly nonlinear function.

In most applications estimating the transfer function takes the most time. When this is done it is often not worth the effort to do the Monte Carlo simulations. Of course there exist exceptions. One such example is when the input distribution is known and it is not a normal distribution.

8.4 Highly Nonlinear Transfer Function

In the case where the transfer function is highly nonlinear, the linearization method does not work. One example of that is the angle error that can arise in a measurement of the length of a rectangle. If the true length is L and the measurement length is $\tilde{L}$ due to the angle error, θ, $\tilde{L}$ will always be larger than L:

$$L = \tilde{L} \cdot \cos(\theta). \tag{8.8}$$

Using the linearization method the sensitivity coefficient for the angle is

$$c_\theta = -\tilde{L} \cdot \sin(\theta), \tag{8.9}$$

and since the expectation for the angle error can be assumed to be zero this leads to $c_\theta = 0$, i.e. the angle error does not contribute to the variance of the length measurement. In this case the Monte Carlo simulations can help. Assume that the angle error is normally distributed:

$$\theta \sim N(0, \sigma^2). \tag{8.10}$$

By simulating a number of normally distributed angles, the variance of the length can be estimated. Here 100,000 simulations were used. The result from the simulations is shown in Table 8.9. The estimated standard deviation for the angle error for different value on σ is shown.

Another example of when the linearization method does not work very well is when boundary conditions make the transfer function quite nonlinear. This happens if the uncertainty is large compared with the limit of the conditions, which should be exceptional in component design.

Table 8.9   Estimated standard deviation for the
angle error assuming different values for σ.

σ	$std(\cos(\theta))$
1	0.0002
5	0.0053
10	0.0211

8.5 Total Variation for Logarithmic Life

In the industrial example presented previously it was necessary to analyse the four different
sources of uncertainties separately. But what is of interest is the total variation of life. Therefore
the four different sources have to be considered together in some way. For the linearization
method this is straightforward. The different variances and covariances are added together. In
the cases where the different sources are independent the total variance is only a summation
of the variances.

In Table 8.10 the standard deviation together with the τ_i for the different sources are shown
using g_1 as a transfer function. From Table 8.10 the total standard deviation for the logarithmic
life is

$$std_{Total}(\ln(N)) = \sqrt{\sum_{j=1}^{4} Var_j(\ln(N))} = 0.16 \qquad (8.11)$$

where j is index of the source. The different sources are assumed to be independent.

In Figure 8.6 the density function for the life is shown for each source separately together
with the variation for the total life. The logarithm of the life is assumed to be normally
distributed. The mean of the logarithmic life is taken as 4.40, which is the logarithm of the life
from the FEM calculations using the nominal input variable. The standard deviation depends
on the source, see Table 8.10, and for the total variation the standard deviation in Equation

Table 8.10   The τ_i for each parameter i and the total standard deviation
for the logarithmic life for each source using g_1 as a transfer function.

Parameter	Geometry τ_i	LCF τ_i	Physical τ_i	Thermal τ_i
x_1	0.062	0.005	0.004	0.0008
x_2	0.058	0.035	0.006	0.021
x_3	0.011	0.048	0.0002	0.105
x_4	0.0007	–	0.026	0.033
x_5	0.003	–	–	–
x_6	0.012	–	–	–
$\hat{std}(\ln(N))$	0.086	0.060	0.027	0.112

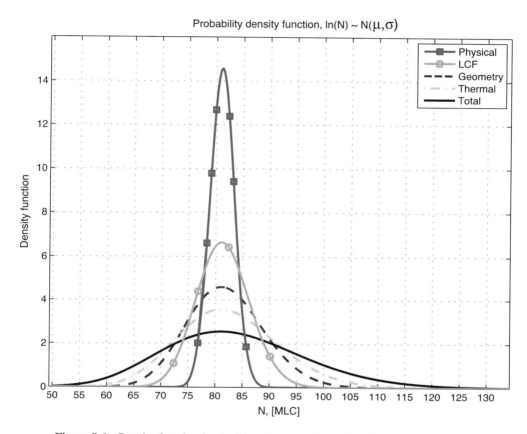

Figure 8.6 Density function for the life using normality and the linearization method.

(8.11) is used. According to Table 8.10 the source that dominates the variation is the *thermal* one and the variation from *physical* sources can almost be neglected.

In Karlén (2007) the probability of being over 50 MLC was estimated, 50 MLC is a minimum criterium for the life. Assuming normality and that the standard deviation is 0.16 and the mean is 4.40 for the logarithmic life, the probability of being over 50 MLC is 0.999.

For the Monte Carlo simulations it is not obvious how to estimate the total variation in life since all sources are analysed separately. In Karlén (2007) two different ways of combining the different transfer functions to one function were used. The Monte Carlo simulations were done on the combined transfer function, which consists of 17 variables. The probability of being over 50 MLC was 0.998 or 0.997 depending on the sampling method and the number of simulations. These probabilities are very close to the probability calculated from the linearization method. Therefore in this example (as in many other practical examples) it is not worth the extra effort of doing a Monte Carlo simulation. After the transfer function is estimated it is enough to perform a weighted summation and use normal distribution assumptions on the output variable.

8.6 Conclusions

The uncertainty in an output variable has been determined using two different methods. The result from the linearization method and the Monte Carlo simulations in this study is almost the same. The distributions of the input variables are often unknown and have to be guessed in some intelligent way. For the Monte Carlo simulation these guesses influence the output distribution a great deal, especially the tail of the output distribution. For the sensitivity case only the second moment of the input variables is used. Therefore it is not necessary to know the complete input distribution.

The form of result from the two methods differs in the same way. For the Monte Carlo simulation the complete output distribution is estimated and for the linearization method only the variance is estimated. It is an advantage that the complete output distribution is estimated, but a disadvantage is that it is easy to think that it is really the correct distribution and thereby believe too much in the tail of the distribution. The tail depends very much on the chosen input distribution and approximate transfer function. Therefore if some of them are wrong the tail can be very wrong. For the linearization method it is more obvious that it is an approximative method. It uses both linearization and normal approximations.

The computer time is greater for the Monte Carlo simulations than for the linearization method. After the transfer function is estimated most of the work has been completed for the linearization method. What is left is a weighted summation of the coefficients in the transfer function, whereas for the Monte Carlo simulation the whole simulation is left.

There are some cases when the two methods give exactly the same result and that is when the transfer function is linear and normal input distributions are used. In this case it is obviously no use doing the simulations. It is enough to estimate the transfer function.

To be able to decrease the variation in the output variable it is important to know which source and variable influence the output variable the most. Here the most important thing is to decrease variation. For the linearization method it is not only possible to identify the source but also which variables in the source influence the variation in the output variable most.

In many applications the transfer function $f(x)$ is not explicitly available and an estimation of $f(x)$ may require more evaluations than is possible to make. The linearization method is based on sensitivity coefficients and for that only the partial derivatives of the function are needed. These can be calculated using difference quotients and that requires fewer evaluations of $f(x)$ than if $f(x)$ needs to be estimated. For each derivative one or two evaluations are needed depending on which type of difference quotient is used, forward, backward or central difference approach. In the difference quotient the step may be chosen in proportion to the standard deviation of the input variable (see Svensson and de Maré, 2008).

The different input variables have different variances and have to be estimated. In some applications it may be possible to decrease the variance for some of the input variables, e.g. by doing more experiments, using better models or through gaining more knowledge. For the Monte Carlo simulations the whole simulations must be done again to see how this affects the variance of the output. For the linearization method only the weighted summation needs to be recalculated. This method is therefore a very flexible one for handling new information that appears.

References

Johansson, P., Chakhunashvili, A., Barone, S. and Bergman, B. Variation mode and effect analysis: a practical tool for quality Improvement. *Quality and Reliability Engineering International*, **22**: 865–876, 2006.

Karlén, K. *Probabilistic design used for mechanical analysis of a rocket nozzle cooling channel*. Master's thesis, KTH, Stockholm, 2007.

Morrison, J. S. Variance synthesis revisited. *Quality Engineering*, **11**: 149–155, 1998.

Svensson, T. and de Maré, J. On the choice of difference quotients for evaluating prediction intervals. *Measurement*, **41**: 755–762, 2008.

Part Three

Modelling

An essential tool in reliability engineering is mathematical models. Their role is to make the different factors causing failures visible and understandable. They then give important insights into the process of constructing counter measures to unreliability.

The construction of good mathematical models is an art rather than a science but insights into some general principles and their applications are helpful for the practitioner. A problem to handle in modelling is the complexity of the model. A simple model is usually seen as an overidealization, which will be of little help. On the other hand, a complex model will include too many unknown entities to be estimated to be able to deliver predictions to be trusted. Chapter 9 supplies an ingenious theory for handling the dilemma between oversimplification and elephantiasis. Chapter 10 uses the theory of Chapter 9 to handle the different sources of variation in a production development process. Chapter 11 focuses on how delicate modelling of variation helps to find new ways to obtain a robust design and Chapter 12 demonstrates that the nonstationarity in load processes generates an extra variation in life. Finally, in Chapter 13, it is noted that there are engineering perspectives, as well as organizational or cultural perspectives, on reliability. A short overview on these aspects is presented.

Robust Design Methodology for Reliability: Exploring the Effects of Variation and Uncertainty
edited by B. Bergman, J. de Maré, S. Lorén, T. Svensson
© 2009, John Wiley & Sons, Ltd

9

Model Complexity Versus Scatter in Fatigue

Thomas Svensson

9.1 Introduction

The fatigue phenomenon is important to understand, both for the purpose of developing materials and for the purpose of designing new constructions based on existing material and component specifications. The first purpose motivates investigations in great detail and it is important to understand the influence of any material property that can be modified in the manufacturing process. In contrast, the second purpose has other demands, and complex models are useful only to the extent of the knowledge of input parameters. Here we will focus on this second purpose, which leads to the problem of prediction of fatigue life for a construction based on material, component, or structure experimental data.

In practice it appears that this problem is solved in industry primarily by using simple models such as the Wöhler curve, the Coffin–Manson relationship, the concept of the fatigue limit, Paris' law, and the Palmgren–Miner cumulative fatigue damage hypothesis; see for instance the review article Berger et al. (2002) and the papers Conle and Chu (1997) and Fermér and Svensson (2001) representing the automotive industry, the aerospace review of Schijve (2003), and Maddox and Razmjoo (2001) dealing with welds. Also many recently developed methods consist of minor modifications of these simple models, see for instance Paris et al. (1999), Taylor (2001), and Wang (1999). In spite of the extensive development of finite element methods in engineering practice, the simple empirical methods still dominate; advanced numerical methods are mostly used for the determination of local stresses.

This chapter is based on the article by T. Svensson, 'Model complexity versus scatter in fatigue', *Fatigue and Fracture of Engineering Materials and Structures*, Vol. 27, pp. 981–990, 2004, and is reproduced by the permission of the publisher Wiley-Blackwell.

Robust Design Methodology for Reliability: Exploring the Effects of Variation and Uncertainty
edited by B. Bergman, J. de Maré, S. Lorén, T. Svensson
© 2009, John Wiley & Sons, Ltd

One explanation to this seemingly old-fashioned industrial practice could be that the information available about the actual material, the actual geometry, and the actual load is not good enough for evaluating the more complex models that are developed in research laboratories. The random scatter is so large that subtle phenomena cannot be resolved within the limits of resources available.

The large scatter in fatigue may be seen as a result of the fact that fatigue is a local phenomenon, i.e. the initiation of the crack that will eventually cause failure depends on *local conditions*, such as microstructural inhomogeneities (triple points, vacancies, interstitial atoms, . . .), grain structure (grain size, orientation, number of slip planes), microscale defects (inclusions, scratches) and milliscale defects (voids, nodules, notches) (Miller, 1999). All these features are important in three dimensions. The crack initiation dependency of these local conditions is extremely complex and no physical models exist that can take them all into account. Still, the sensitivity of the local condition is large and the assessment of the most critical set of local conditions is crucial for a good fatigue life prediction.

Even if there were good physical models for the crack development from a certain set of local conditions, there would be no possibility for the engineer to examine the actual component with respect to all microstructural features. At best, he can rely on specifications such as minimum notch radii or surface roughness. Materials are sometimes specified with respect to mean grain size, mean inclusion size, overall chemical composition and similar things, but the local variations of these features are unknown to the engineer.

Two approaches for modelling can be identified: (1) Use detailed experiments to find a model close to the physical reality which contains a large number of parameters. Delete the parameters that are overwhelmed by scatter or are unmeasurable and replace them by their expected or nominal values. (2) Start with the simple empirical models already in use, i.e. the Wöhler curve, Paris' law, the Coffin–Manson formula and the Palmgren–Miner rule and find sequentially new measurable parameters that diminish scatter but whose influence is not overwhelmed by scatter. Both approaches are useful, but in both cases there is a need to quantify the judgement 'overwhelmed by scatter'. This problem leads, for instance, to stepwise regression solutions but has rarely been addressed in the fatigue community, although it is investigated in, for instance, Wang (1999), Kozin and Bogdanoff (1989), Cremona (1996) and Svensson (1997). Of course, the problem is not specific to fatigue, but is a general problem of physical modelling and it has been analysed by different authors, for instance in Juditsky et al. (1995) and Akaike (1974) for time series problems and in LuValle and Copeland (1998) for accelerated testing strategies.

Breiman and Freedman (1983) have given a mathematical formulation of the problem that is suitable for our purposes. An evaluation of their idea in the fatigue context provides hints as how to approach the problem of life prediction model complexity. This chapter first presents the Breiman formulation and results, then evaluates a criterion for adding a variable to a model, and finally uses a unified fatigue model for appreciation of the optimal complexity.

9.2 A Statistical Model

In order to study the trade-off between complexity and parameter knowledge we follow the formulation given by Breiman and Freedman (1983). A general statistical model for the logarithm of fatigue life can be approximated by a linear function:

$$Y = \sum_{i=1}^{M} \beta_i X_i + \varepsilon, \tag{9.1}$$

where Y is the logarithm of the fatigue life, X_i are variables in the model and ε is a random error, assumed to be independent between different trials and with a common variance, $Var[\varepsilon] = \sigma^2$. The model parameters are represented by β_i. Once such a model has been established, certain values of the variables X_i are chosen for a number of experiments (in this chapter denoted n) and these together with the resulting Y values are used to estimate the model parameters β_i by linear regression. It is clear that the more parameters that are introduced, the better fit one can obtain for the measured response. However, if the number of tests is fixed, the uncertainty of the parameter estimates will increase when the number of parameters increases. This effect can be quantified in quite a general way for linear functions such as Equation (9.1), and here we will use such a quantification to investigate the role of scatter and modelling in fatigue.

In order to obtain a simple expression for the prediction variance, the variables X_i are modelled as random properties. This means that one looks at the population of parts, specimens or constructions and regards the variables as randomly distributed in this population. In addition, in order to obtain simple expressions, we assume that they are normally distributed.

As an example from fatigue analysis, we have the empirical model of Basquin

$$N = \alpha \Delta S^{-\beta}, \tag{9.2}$$

where N is the number of cycles to failure at the load range ΔS, and α and β are model parameters. After a log transformation the Basquin equation reads

$$\ln N = \ln \alpha - \beta \ln \Delta S.$$

The statistical model corresponding to this transformed Basquin equation is

$$\ln N = \ln \alpha - \beta \ln \Delta S + e, \tag{9.3}$$

and thereby fits in the general model (Equation 9.1) with $M = 2$ by defining $\ln N = Y$, $1 = X_1$, $-\ln \Delta S = X_2$, $\ln \alpha = \beta_1$, $\beta = \beta_2$ and $e = \varepsilon$. In the physical world the observed randomness is partly explained by hidden deterministic errors and the random error e in Equation (9.3) may be modelled as

$$e = \sum_{i=3}^{M} \beta_i X_i + \varepsilon,$$

where the $\{X_i; i > 2\}$ are unknown or neglected variables interpreted to be random choices among the population of entities. By this formulation, the general linear model (Equation 9.1) can be rewritten as

$$Y = \sum_{i=1}^{p} \beta_i X_i + \sum_{j=p+1}^{M} \beta_j X_j + \varepsilon, \tag{9.4}$$

where p is the number of variables used in the model. When the p parameters in such a model are estimated by ordinary linear regression techniques, one also obtains an estimate of the error variance, which in reality will be an estimate of the sum of two variances, namely $\sigma_p^2 + \sigma^2$,

$$\sigma_p^2 = Var\left[\sum_{j=p+1}^{M} \beta_j X_j \,|\, X_1 \ldots X_p\right], \quad \sigma^2 = Var[\varepsilon]. \tag{9.5}$$

Here two variance components are introduced, one representing the $M - p$ ignored variables conditioned on the values of the variables used in the regression, and the other representing the remaining random error, which is assumed to be independent of all M variables.

9.2.1 Prediction Uncertainty

In engineering fatigue design, one wants to predict the fatigue life for a certain material, component or structure. Based on laboratory tests, the engineer then uses a set of estimated parameters $\{\hat{\beta}_1(p), \hat{\beta}_2(p), \ldots, \hat{\beta}_p(p)\}$ and a set of variable values $\{X_1, X_2, \ldots X_p\}$ to predict the fatigue life,

$$Y^* = \sum_{i=1}^{p} \hat{\beta}_i(p) X_i. \tag{9.6}$$

Note that the regression coefficients depend on p, i.e. on the number of parameters used, since there might be dependencies among the variables and for the parameters in Equation (9.4) $\beta_i = \beta_i(M)$ (see Breiman and Freedman (1983) for details). The prediction error is then $Y - Y^*$ and the variance of this property is

$$Var\left[Y - Y^*\right] = E_X\left[Var\left[Y - Y^*\right] | X_1, X_2, \ldots, X_p\right].$$

Since this variance is taken as the expectation over the random X space, it depends only on the number of included variables p and the number of reference tests n, and is therefore denoted $U_{n,p}$. The important result given in Breiman and Freedman (1983) is the following: If the parameter estimation has been done in accordance with Equation (9.4), then the mean prediction error variance is given by

$$U_{n,p} = \left(\sigma^2 + \sigma_p^2\right)\left(1 + \frac{p}{n-1-p}\right), \quad p \leq n - 2, \tag{9.7}$$

where n is the number of tests in the estimation stage. In this formula the first term within the second parenthesis represents the variance of the future observation and the second term reflects the uncertainty of the estimated model parameters. Note that $U_{n,p}$ is the mean variance with respect to the random X variables and that it is assumed that the same population of X variables is used both in the estimation procedure and in the prediction stage, i.e. the n tests should represent the same population of variable values as the population whose life one wants to predict. This condition is often not fulfilled, which results in systematic errors not included in (9.7), see Section 9.2.3 for more comments about this.

From Equations (9.5) and (9.7) it can be seen that, if the number of regression variables increases, σ_p^2 decreases, but the factor $(1 + p/(n - 1 - p))$ increases. This means that there exists an optimal choice of the number of variables and the result (Equation 9.7) can be seen as a mathematical formulation of the trade-off between model complexity and parameter knowledge.

9.2.2 Conditions for Increasing the Complexity

Using Equation (9.7) one can calculate conditions for increasing complexity. Assume that a model with p parameters is given and we want to find the condition for adding one parameter. Assume first that

$$Var[\beta_{p+1}(p+1)X_{p+1}|X_1 \ldots X_p] = c^2(\sigma^2 + \sigma_p^2), \qquad 0 < c \leq 1 \qquad (9.8)$$

i.e. the conditional variance of the $(p+1)$-th variable is a fraction of the error variance using p variables and that the values of the variable X_{p+1} are known both at the testing stage and at the prediction stage. We can then derive the condition for adding the variable, which is that the prediction variance should decrease:

$$U_{n,p+1} < U_{n,p} \qquad (9.9)$$

The error variance $U_{n,p}$ is given above (Equation 9.7) but in order to formulate $U_{n,p+1}$ we need an expression for σ_{p+1}^2. Actually, from the regression theory (Breiman and Freedman, 1983) one can conclude that

$$\sigma_{p+1}^2 = \sigma_p^2 - Var[\beta_{p+1}(p+1)X_{p+1}|X_1 \ldots X_p], \qquad (9.10)$$

where $\beta_{p+1}(p+1)$ is the coefficient obtained from regression of Y on $\{X_1, X_2, \ldots, X_{p+1}\}$. The condition (Equation 9.9) can then be written

$$(\sigma^2 + \sigma_p^2 - c^2(\sigma^2 + \sigma_p^2))\left(1 + \frac{p+1}{n-2-p}\right) < (\sigma^2 + \sigma_p^2)\left(1 + \frac{p}{n-1-p}\right),$$

which can be simplified to the solution

$$c^2 > \frac{1}{n-p-1}, \qquad (9.11)$$

i.e. the conditional standard deviation of the entity $\beta_{p+1}(p+1)X_{p+1}$ should be at least $\sqrt{\frac{\sigma^2 + \sigma_p^2}{n-p-1}}$. In Figure 9.1, the result is illustrated for three different values of n. It is seen that if 30 tests are used for model identification, the new variable should be added only if its standard deviation influence is at least approximately 20% of the standard deviation of the smaller model.

Note that the variance property in question is the conditional variance, which means that it is the variance for the new variable when all the others are kept constant. This means that if the new variable is dependent on some of the other p variables, the standard deviation to be compared is only a part of the unconditional variable variance. For instance, in fatigue many material variables are strongly dependent, such as yield stress, tensile strength and mean grain size, and in such cases the nonconditional standard deviation must be even larger to fulfil the condition.

By sorting the variables by decreasing variability, one can check for the fulfilment of the criterion for one variable at a time. For each variable included, the number of variables will increase to $p+1$ and the criterion on c will be strengthened by Equation (9.11). On the other

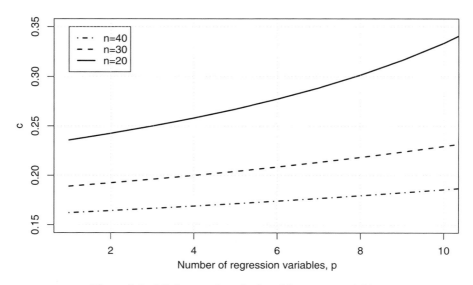

Figure 9.1   Minimum value of c for adding a new variable.

hand the remaining variance will decrease from $\sigma^2 + \sigma_p^2$ to $\sigma^2 + \sigma_{p+1}^2$ so the proportionality constant c must be recalculated.

It may be seen to be quite difficult to estimate the conditional variance for a new variable. However, in a specific situation, when one has performed n measurements, it is easy to apply the criterion by using the results from two regressions. Namely, from Equations (9.10) and (9.8) one obtains

$$c^2 = \frac{Var[\beta_{p+1}(p+1)X_{p+1} \,|\, X_1 \ldots X_p]}{(\sigma^2 + \sigma_p^2)} = \frac{\sigma_p^2 - \sigma_{p+1}^2}{(\sigma^2 + \sigma_p^2)}$$

$$= \frac{(\sigma^2 + \sigma_p^2) - (\sigma^2 + \sigma_{p+1}^2)}{(\sigma^2 + \sigma_p^2)}$$

$$= 1 - \frac{(\sigma^2 + \sigma_{p+1}^2)}{(\sigma^2 + \sigma_p^2)} \approx 1 - \frac{S_{p+1}^2}{S_p^2},$$

where S_p^2 and S_{p+1}^2 are the mean square errors in two regressions, one with p variables and the other with $p + 1$ variables included,

$$S_p^2 = \frac{1}{n-p} \sum_{j=1}^{n} (Y_j - \widehat{Y}_j)^2, \quad \widehat{Y}_j = \sum_{i=1}^{p} \widehat{\beta}_i X_{ij}.$$

9.2.2.1 Not Precisely Measurable Variables

Knowledge about the variables is often not as good in the prediction situation as under the laboratory test conditions. Specimen geometry, for instance, is better known than structure or component geometry. Furthermore, the stress in a component in service may be very uncertain both with regard to the global stress that depends on its future usage, and with regard to the distribution of stress to the critical location. These uncertainties are important for the trade-off between complexity and knowledge and we will first extend the previous result for the case when the new added variable is uncertain at the prediction stage. Therefore, assume next that at the prediction stage the new variable is not known exactly, but we still want to use the estimated coefficients from the laboratory test result. This situation may be modelled by adding an uncertainty random variable Z to X_{p+1} at the prediction stage, i.e.

$$\widehat{Y}^* = \sum_{i=1}^{p} \widehat{\beta}_i (p+1) X_i + \widehat{\beta}_{p+1} (p+1) \left(X_{p+1} + Z \right) = \sum_{i=1}^{p+1} \widehat{\beta}_i (p+1) X_i + \widehat{\beta}_{p+1} (p+1) Z.$$

By assuming that the random variable Z is independent of all the included X variables, the prediction error variance will increase by the variance of the extra term $\widehat{\beta}_{p+1}(p+1)Z$. To see the influence of this lack of knowledge we assume that this extra variance is a certain proportion of the variance of the new variable. With the proportionality constant a^2 we obtain

$$Var\left[\widehat{\beta}_{p+1} (p+1) Z \right] = a^2 c^2 \left(\sigma^2 + \sigma_p^2 \right).$$

The condition is then

$$\left(\sigma^2 + \sigma_p^2 - c^2 \left(\sigma^2 + \sigma_p^2 \right) \right) \left(1 + \frac{p+1}{n-2-p} \right) + a^2 c^2 \left(\sigma^2 + \sigma_p^2 \right)$$
$$< \left(\sigma^2 + \sigma_p^2 \right) \left(1 + \frac{p}{n-1-p} \right).$$

For $a^2 < \frac{n-1}{n-2-p}$ we obtain the solution

$$c^2 > \frac{n-1}{(n-p-1)\left[n-1-a^2(n-2-p)\right]} \tag{9.12}$$

This condition is illustrated in Figure 9.2 for different values of a for the case when $n = 30$ and $p = 2$. It is seen in the figure that uncertainties in the variable value increase the demands on the relative information in the new variable, but the increase may be judged to be unexpectedly small. The reason for this is that the precision of the p first parameter estimates will be improved by reducing the error term regardless of the uncertainty in the prediction situation. This is actually always the case, even when this second type of condition is not fulfilled. Therefore, the criterion assuming full knowledge about the variable values (Equation 9.11) should always be used in the estimation stage. However, if the second criterion (Equation 9.12) is not fulfilled the new variable should not be used at the prediction stage. However, this change of parameter set will usually make it necessary to recalculate the parameters remaining in the model.

Another problem arises when a variable with prediction uncertainty has fulfilled the criterion and is thereby included in the model. Then the inclusion of additional variables may fulfill the criteria, but the decrease in prediction uncertainty is negligible in comparison with the total

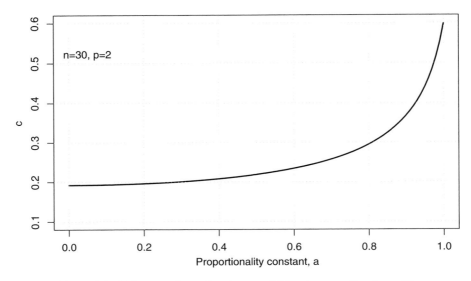

Figure 9.2 Influence of a random input variable when $p = 2$ and $n = 30$.

uncertainty. This is typically the case when the prediction uncertainty of the actual stress is large. In such cases it may be more efficient to use resources on better stress analyses than performing detailed analyses of, for instance, initial crack sizes.

9.2.3 Bias

Equation (9.1) is also illustrative for systematic model errors, i.e. bias in the life prediction. Bias in the prediction situation will be present if some of the ignored variables $\{X_{p+1} \ldots X_M\}$ are not representative for the prediction situation. For instance, if a Basquin relationship is established for the load ratio $R = \frac{S_{min}}{S_{max}} = 0$, it will give biased predictions for the load ratio $R = -1$. This happens because the the X variable representing the load ratio is ignored in the model and its values are not randomly chosen either in the laboratory test, nor in the prediction situation. If, on the other hand, the laboratory test was performed at different R ratios, representing the same population as the prediction case, then there would not be any bias, but instead the random error term $\sum_{j=p+1}^{M} \beta_j X_j + \varepsilon$ would increase. Large but correct confidence limits would be the result. This shows that it is important to have knowledge about the ignored variables in order to avoid bias and make relevant reference tests.

9.3 Design Concepts

Our aim is to discuss the complexity problem in fatigue life prediction. The prediction procedure depends substantially on the application and our analysis must take these differences into consideration. The following design concepts in fatigue can be identified and the choice between them may be determined by either safety requirements, practical considerations or the actual required life:

1. Design for finite life, low cycle fatigue. High stresses just above the yield stress create cracks quite quickly. Design is usually based on the empirical Coffin–Manson relationship consisting of a linear fit to the log–log relationship between plastic strain and life.
2. Design for infinite life. The safest design concept is to avoid the possibility of fatigue failure, i.e. to design against the fatigue limit. The very possibility of defining a fatigue limit with regard to infinite life is debatable, but in practice the fatigue limit may be interpreted as an endurance limit; the load range that will cause no fatigue failure during a very high number of cycles, given by the expected life for the actual product. Designing against an endurance limit does not include a life prediction, but empirically estimates a stress level that will not cause any crack to grow to failure. This design concept may be seen as the result of too large a prediction uncertainty in life prediction.
3. Design for finite life, including both initiation and propagation. For parts with smooth surfaces and without sharp notches the major part of the damage accumulation is the so-called crack initiation process. This may be seen as a combination of both true initiation and growth of microstructural small cracks. Design is mainly based on the empirical Wöhler curve consisting of a linear fit to the log–log relationship between stress range and life. The scatter around the Wöhler curve is often as large as a factor of five (NRIM, 1979).
4. Design for finite life, propagation only. For welded parts or parts with rough surfaces or sharp notches the crack propagation part of the fatigue process dominates and thereby the scatter diminishes. Also here the dominating design method seems to be a Wöhler curve, possibly derived from crack growth curves, but scatter may be reduced to a factor of two (Schijve, 1994).
5. Damage tolerance design. For parts where cracks may occur but with high safety requirements it is necessary to design against crack growth. Regular inspections guarantee certain maximum crack lengths and the remaining fatigue life can be predicted by means of the linear fracture mechanics concept. One usually uses an empirical relationship between the crack growth rate and the stress intensity factor. The random scatter around the derived life is here about a factor of two (Virkler et al., 1979).

The aim of our further study is to investigate which variables it is reasonable to include in fatigue models. We will do this for the three last design concepts given above by starting with a minimum number of variables and investigate the result of adding new variables in light of the criteria discussed above. To this end one must first estimate the initial property $\sigma_{p_0}^2 + \sigma^2$ for different design concepts. Here p_0 is the minimum number of parameters. As fatigue life is often studied in a log transformation, one would prefer to measure scatter as the *coefficient of variation*, and we define this property for a positive random variable X as the relative standard deviation,

$$w_X = \frac{\sigma_X}{E[X]}.$$

Table 9.1   Approximate Coefficients of Variations
of the Scatter.

Initiation and propagation	1
Propagation	0.5
Damage tolerance	0.7

In our case we then need estimates of

$$w_N = \frac{\sqrt{\sigma_{p_0}^2 + \sigma^2}}{E[N]}.$$

For all design concepts we will start with the stress as the first included variable and also regard this as the minimum choice of variables since the mean behaviour must also be included – represented in the Wöhler curve by the constant α, $p_0 = 2$. In the case of design against initiation and propagation, w_N then corresponds to the scatter in the ordinary Wöhler curve. A reasonable general estimate of this scatter can be set to 100%, $w_N = 1$. In the case of propagation only, i.e. fatigue for parts that are not polished, again the ordinary Wöhler curve shows the scatter. In this case we assume that the scatter decreases to $w_N = 0.5$. In the case of damage tolerance design both initial crack size and a geometry variable are included in engineering practice, but here we start with stress only. This means that the initial scatter will include the variation of initial crack lengths and geometry in the actual population. These variations will of course depend on the application and here we just make the rough guess that $w_N = 0.7$ (Table 9.1).

9.4 A Crack Growth Model

In order to find reasonable estimates of the coefficients β_i in fatigue, we need a model that includes the most important variables and simple enough to allow a rough analysis of the general behaviour. One such model is a semi-empirical crack growth equation presented by McEvily et al. (2003), dealing with the whole damage development, including short crack growth:

$$\frac{da}{dN} = A\left[\left(\sqrt{2\pi r_e}F + Y\sqrt{\pi a}F\right)\Delta\sigma - \left(1 - e^{-k(a-a_0)}\right)\left(K_{op,\max} - K_{\min}\right) - \Delta K_{eff,th}\right]^2,$$

(9.13)

where A is a variable depending on material and environment, a is the crack length, and $\Delta\sigma$ is the stress range. The parenthesis preceding the stress range is an expression including a modification of the stress intensity factor for short cracks. Here r_e is considered to be the size of an inherent flaw, F is a correction factor for plasticity given by $F = \frac{1}{2}(\sec\frac{\pi}{2}\frac{\sigma_{\max}}{\sigma_Y} + 1)$, and Y depends on the crack or flaw geometry. The second term within the main parenthesis is present to subtract the part of the stress intensity range that is not harmful. It consists of two parts, one modelling crack closure, where the exponential factor accounts for the development of closure when a crack advances. Here k is a material dependent variable and a_0 is the initial crack (or flaw) size. The entity $K_{op,\max}$ is the stabilized crack opening stress intensity when the crack is large enough, $K_{\min}$ is the minimum stress intensity and $\Delta K_{eff,th}$ is the intrinsic threshold for the material.

The model can be seen as an attempt to include the knowledge about short crack behaviour, crack closure and threshold effects in the traditional Paris' law, but keeping it as simple as possible. Using the model for different steels with different defect sizes shows qualitatively strong agreement with observations regarding the crack growth behaviour (McEvily et al., 2003). Therefore, it is suitable for studying the influence of different variables on the fatigue life prediction. The model can be subjected to numerical integration and the result may be written as a function,

$$N = f(X_1, X_2, \ldots, X_{10}) = f\left(a_0, a_C, \Delta\sigma, A, Y, \Delta\sigma_{w0}, \sigma_Y, k, \Delta K_{eff,th}, K_{op,\max}, K_{\min}\right),$$

$$(9.14)$$

where a_0 and a_C are the initial and final crack lengths, respectively, and the variable r_e has been substituted by the fatigue limit $\Delta\sigma_{w0}$ through the identity (McEvily et al., 2003)

$$\Delta K_{eff,th} = \Delta\sigma_{w0}\left(\sqrt{2\pi r_e F} + Y\sqrt{\pi r_e F}\right).$$

9.4.1 Gauss Approximation Formula

The model of Equation (9.13) is not linear in its variables, not even after log transformation. In order to apply the results above, we will use the Gauss approximation formula. Referring to Equation (9.14) it reads

$$\sigma_N^2 \approx \sum_{i=1}^{10} \left(\frac{\partial N}{\partial X_i}\right)^2 \sigma_{X_i}^2 + \sum_{i \neq j} \frac{\partial N}{\partial X_i} \frac{\partial N}{\partial X_j} Cov\left(X_i, X_j\right) \tag{9.15}$$

For simplicity, we will first drop the covariance terms and assume that the variables are independent. Since a log transformation is probably more linear, one would prefer studying the squared coefficients of variation in place of the variances. The approximation (Equation 9.15) will then change to

$$w_N^2 \approx \sum_{i=1}^{10} c_i^2 w_{X_i}^2, \qquad c_i = \left|\frac{\partial N}{\partial X_i}\right|_{X_i=E[X_i]} \frac{E[X_i]}{E[N]}\right|, \tag{9.16}$$

where c_i will be referred to as the sensitivity coefficient.

9.4.2 Sources of Variation

Numerical calculations were carried according to Equation (9.14) for a mild steel, using parameters and variables given in McEvily et al. (2003). All calculations used the constant load ratio $R = -1$, which means that $K_{\min} = -\frac{\Delta K}{2}$, and this variable can be dropped from Equation (9.14), resulting in ten variables. Two situations are simulated, first for the development of a crack from 200 μm to 5 mm. The evaluation is made at load ranges chosen to give ten million, one million and one hundred thousand cycles respectively. The resulting sensitivity coefficients

Table 9.2 Sensitivity coefficients for crack growth from 0.2 mm to 5 mm.

$\Delta\sigma_{nom}$	N_{nom}	c_{a_0}	c_{a_C}	$c_{\Delta\sigma}$	c_A	c_Y	$c_{\Delta\sigma_{w0}}$	c_{σ_Y}	c_k	$c_{\Delta K_{eff,th}}$	$c_{K_{op,max}}$
286	10^7	2.9	0.03	27	1.0	16	7.3	1.8	5.1	6.3	9.6
340	10^6	0.7	0.1	8.1	1.0	4.5	2.2	1.8	1.4	0.91	1.0
474	10^5	0.3	0.3	7.7	1.0	2.3	0.8	4.4	0.3	0.04	0.5

can be seen in Table 9.2. Similar calculations were carried out for the crack propagation phase by using a 1 mm initial crack size. These results are shown in Table 9.3.

From the tables it can be seen that the sensitivity for the load range is very high, especially at long lives. Note that this sensitivity coefficient can be interpreted as the Wöhler exponent, since in the case of the Basquin Equation (9.2),

$$c_{\Delta S} = \left| \frac{\partial N}{\partial \Delta S} \right|_{\Delta S = E[\Delta S]} \frac{E\,[\Delta S]}{E\,[N]} = \left| -\alpha\beta E\,[\Delta S]^{-\beta-1} \frac{E\,[\Delta S]}{E\,[N]} \right| = \beta$$

It can also be seen that the crack geometry (Y) is highly influential and so are the maximum opening stress intensity factor ($K_{op,max}$), the effective threshold ($\Delta K_{eff,th}$), the delay coefficient (k) and the initial crack size (a_0). Also the fatigue limit ($\Delta\sigma_{w0}$) is important, but it is highly correlated with the initial crack length and must be analysed in more detail with this fact in mind. Here, we neglect this variable for simplicity and regard its influence as essentially contained in the initial crack length. Most sensitivity coefficients decrease when the predicted life decreases, which is in accordance with the common observation that the scatter in the Wöhler curve usually increases at long lives.

The other factors that affect the total variance are the coefficients of variation of the influential variables, $\{w_i; i = 1 \ldots 10\}$. These should reflect the variation of the different variables within a population of parts that are subject to the fatigue design. Of course, these values are specific for each application and can be changed by narrowing the application. Here, we use quite rough guesses of these coefficients (Table 9.4), but try to emphasize the differences between design concepts.

9.4.3 Obtaining Proper Model Complexity

We will here apply Equation (9.11) to the three different fatigue design concepts. Since the values in Tables 9.1–9.4 are quite rough estimates, we will also simplify the criterion to $c = 20\%$, This approximately corresponds to 30 reference tests and two to six variables (see Figure 9.1).

Table 9.3 Sensitivity coefficients for crack growth from 1 mm to 5 mm.

$\Delta\sigma_{nom}$	N_{nom}	c_{a_0}	c_{a_C}	$c_{\Delta\sigma}$	c_A	c_Y	$c_{\Delta\sigma_{w0}}$	c_{σ_Y}	c_k	$c_{\Delta K_{eff,th}}$	$c_{K_{op,max}}$
218	10^7	4.6	0.1	20	1.0	15	4.1	0.2	2.1	5.6	7.8
274	10^6	1.3	0.4	6.5	1.0	4.6	1.2	0.6	0.5	1.1	1.7
420	10^5	0.7	0.6	5.4	1.0	2.5	0.6	2.2	0.2	0.1	0.5

Table 9.4 Coefficients of variation at different design concepts (rough guesses).

Application	w_{a_0}	w_{a_C}	w_A	w_Y	w_{σ_Y}	w_k	$w_{\Delta K_{eff,th}}$	$w_{\Delta K_{op,max}}$
Damage tolerance	0.3	0.01	0.05	0.1	0.02	0.1	0.01	0.01
Welds	0.1	0.05	0.05	0.05	0.02	0.1	0.01	0.02
Polished	0.2	0.1	0.05	0.05	0.02	0.1	0.01	0.02

Combining the values in Tables 9.2–9.4 shows that the most essential variables are Y, a_0, k and A. For the initiation and growth case (polished parts) and life 10^6, the effects of Y and a_0 are

$$c_Y w_Y = 4.5 \cdot 0.05 = 0.23 \quad \text{and} \quad c_{a_0} w_{a_0} = 0.7 \cdot 0.2 = 0.14.$$

If we have a scatter in the Wöhler curve equal to 1 and want to add the variable Y, then its effect $c_Y w_Y$ should be at least 20% of $1 = 0.2$, which is actually the case. In accordance with Equation (9.10), after adding this variable the new scatter term will have the coefficient of variation $w_{\varepsilon_2} = \sqrt{1 - 0.23^2} = 0.97$, and the the initial crack length does not fulfil the criterion. The resulting model function should then include load and geometry and neglect all the other variables. It cannot be written explicitly, but a Taylor approximation may give some insight into its behaviour.

A Taylor expansion of the logarithm of the life with respect to the given essential variables can be written

$$\ln N = \ln N_0 + \frac{\partial \ln N}{\partial \ln \Delta\sigma}(\ln \Delta\sigma - \ln \Delta\sigma_0) + \frac{\partial \ln N}{\partial \ln Y}(\ln Y - \ln Y_0) + \varepsilon',$$

where the life is evaluated around a certain reference value N_0, which is expected for the variable values $\Delta\sigma_0$ and Y_0. The error term ε' includes both model errors and random errors, $\varepsilon' = \sum_{j=3}^{10} \beta_j X_j + \varepsilon$. Since $\frac{\partial \ln N}{\partial \ln Y} = \frac{Y}{N}\frac{\partial N}{\partial Y} = c_Y$ the formula can be rewritten as

$$\ln N = \ln N_0 + c_{\Delta\sigma} \ln \frac{\Delta\sigma}{\Delta\sigma_0} + c_Y \ln \frac{Y}{Y_0} + \varepsilon',$$

and taking antilogarithms

$$
\begin{aligned}
N &= N_0 \cdot \left(\frac{Y}{Y_0}\right)^{c_Y} \cdot \left(\frac{\Delta\sigma}{\Delta\sigma_0}\right)^{c_{\Delta\sigma}} \cdot E' \\
&= \left(N_0 Y_0^{-c_Y} \Delta\sigma_0^{-c_{\Delta\sigma}}\right) \cdot Y^{c_Y} \cdot \Delta\sigma^{c_{\Delta\sigma}} \cdot E' \\
&= \alpha_0 \cdot Y^{c_Y} \cdot \Delta\sigma^{c_{\Delta\sigma}} \cdot E'.
\end{aligned}
$$

It may be observed that this formula resembles the Basquin equation, but there are three essential differences. The first one is that the empirical constants in the Basquin equation are replaced by the mean values of the included variables in combination with sensitivity coefficients. The second difference is that the variable Y is included, which implies that crack geometry should be taken into account. The third is that the exponent for the stress range is not constant, but ranges from 7.7 at 10^5 cycles to 27 at 10^7 cycles. However, for the narrower life range 10^5–10^6 it is fairly constant, which is in accordance with the observations of highly uncertain behaviour close to the fatigue limit.

For the case of crack growth in welded parts and life 10^6 the effects of Y and a_0 are

$$c_Y w_Y = 4.6 \cdot 0.05 = 0.23 \quad \text{and} \quad c_{a_0} w_{a_0} = 1.3 \cdot 0.1 = 0.13.$$

If we have a scatter in the Wöhler curve equal to 0.5 and want to add the variable Y, then it should be at least 20% of 0.5, i.e. 0.10, which is the case. After adding this variable, the new scatter term will have the coefficient of variation $w_{\varepsilon_2} = \sqrt{0.5^2 - 0.23^2} = 0.44$. Twenty per cent of $0.44 = 0.09$ and the initial crack variable should also be included. The next two variables in turn are

$$c_k w_k = 1.4 \cdot 0.1 = 0.14 \quad \text{and} \quad c_A w_A = 1.0 \cdot 0.05 = 0.05.$$

$w_{\varepsilon_3} = \sqrt{0.44^2 - 0.14^2} = 0.42$. Twenty percent of 0.42, equals 0.08, and further $w_{\varepsilon_4} = \sqrt{0.42^2 - 0.14^2} = 0.40$. Twenty per cent of 0.40 is 0.08, and the evaluation stops there. The resulting model would include the load, the geometry, the initial crack length and the closure coefficient. However, here the engineer is usually faced with the fact that the knowledge about these parameters in the prediction phase is limited.

For the damage tolerance case and life 10^6, the effects of Y and a_0 are

$$c_Y w_Y = 4.6 \cdot 0.1 = 0.46 \quad \text{and} \quad c_{a_0} w_{a_0} = 1.3 \cdot 0.3 = 0.39.$$

If we have a scatter in the Wöhler curve equal to 0.7 and want to add the variable Y, then it should be at least 20% of 0.7, i.e. 0.14, which indeed is the case. After adding this geometry variable, the new scatter term will have the coefficient of variation $w_{\varepsilon_2} = \sqrt{0.7^2 - 0.46^2} = 0.53$. Twenty per cent of 0.53 is 0.11 and the initial crack length variable should also be included. The next two variables in turn are

$$c_k w_k = 1.4 \cdot 0.1 = 0.14 \quad \text{and} \quad c_A w_A = 1.0 \cdot 0.05 = 0.05.$$

$w_{\varepsilon_3} = \sqrt{0.53^2 - 0.39^2} = 0.36$. Twenty per cent of 0.36 is 0.07, and further $w_{\varepsilon_4} = \sqrt{0.36^2 - 0.14^2} = 0.33$. Twenty per cent of 0.33 is 0.07, and the criterion is not fulfilled. The resulting model for the damage tolerance case would, just like the weld case, include initial crack length, geometry and the closure exponent.

The rough estimates made above are based on our first criterion (Equation 9.11), assuming full knowledge about the variables included. However, in different design concepts this knowledge varies a great deal. In the damage tolerance case the actual parameters are often measurable to quite good accuracy. Initial crack lengths are determined by inspection, crack geometry is well known, and closure characteristics are often well investigated. In consequence this is an area where more complex models than the Wöhler curve are in industrial use. In contrast, weld cracks are not very well known in advance, neither their initial length nor their geometry nor closure characteristics, and more detailed investigations using the second criterion (Equation 9.12) would probably exclude some of these parameters.

It can be concluded from these calculations that the models for the different design concepts are close to the models used in practise, i.e. the Wöhler curve for propagation and initiation problems and Paris' law for damage tolerance design. In the Wöhler curve cases, crack geometry, initial crack length and closure coefficient seem to be candidates for improved models, but their inclusion is highly dependent on the possibility of measuring the variables in question. In the damage tolerance case the three resulting candidates are already used in practice, even if the closure characteristics are usually modelled by some other parameter.

9.5 Partly Measurable Variables

If the Wöhler exponent is 8 and one wants to design against limited life, the Gauss approximation formula (Equation 9.16) applied on the Wöhler curve tells us that approximately

$$w_N^2 \approx \sum c_i^2 w_{X_i}^2 = 64 w_{\Delta\sigma}^2 + \dots$$

This means that if the uncertainty in the load range is, say 5%, it will alone give an uncertainty in life of 40%! Since this precision in load is difficult to obtain when estimating service loads, it is easy to understand that in case of large Wöhler exponents (long initiation part, smooth surfaces) the only reasonable design criterion is the fatigue limit or the endurance limit. In the same way one can often see that an uncertain load prediction makes other sources of variation negligible, and essential improvements of fatigue reliability are highly dependent on methods for obtaining critical stresses. Four serious difficulties arise in stress estimation: (1) customer usage variation; (2) the distribution of global loads to local stresses within a construction; (3) multiaxiality; and (4) variable amplitude. The first problem is clearly identified and mostly unsolved in automotive applications (Berger et al., 2002), where loads from proving grounds appear to be the best possible estimate at present. The second problem can, for instance, be identified in weld applications, where finite element calculations of stresses can show differences of a factor of two between different calculation methods (Pers et al., 1997; Katajamäki et al., 2002; Doerk et al., 2003). The third and fourth problems are connected to the problem of equivalence between laboratory tests and service and are subject to extensive research at present. In fact, the problem of variable amplitude may primarily be a problem of lack of knowledge about the damage driving force and not a problem of too simplistic modelling. This conclusion could be drawn from the experimental results given in Jono (2003).

9.6 Conclusions

The scatter in fatigue life is of the utmost importance for modelling at the design stage. Since only a few experiments can be done in usual engineering practice, the number of variables to be used in models must be severely limited. The rough calculations made in this chapter suggest that the models used in engineering practice are close to the optimum.

In a specific application, the technique evaluated here may be used for the determination of the optimal choice of variables, and judgements can be made about where resources should be directed for better fatigue reliability. Knowledge about the variation of different influential variables is critical for such work, both for the optimal choice of model complexity and for the avoidance of systematic prediction errors. By modelling the variables as random properties, one can see that the same population must be used for the reference set as for the prediction set. A careful analysis of the influential variables is thereby useful even if the variables are not included in the model.

The results suggest that methods for improving fatigue reliability in industry should not primarily consist of more complex models, but should rather concentrate on reducing scatter. This may be done along different lines, for instance through better stress calculations, by service load measurements or by narrowing the applications by choosing a better defined population for reference test and predictions.

References

Akaike, H. New look at the statistical model identification. *IEEE Transactions on Automatic Control*, **AC-19**(6): 716–723, 1974.

Berger, C., Eulitz, K-G, Hauler, P., Kotte, K.-L., Maundorf, H., Schütz, W., Sonsino, C. M., Wimmer, A. and Zenner, H. Betriebsfestigkeit in germany – an overview. *International Journal of Fatigue*, **24**: 603–625, 2002.

Breiman, L. and Freedman, D. How many variables should be entered in a regression equation? *Journal of the American Statistical Association*, **78**(381): 131–136, 1983. Theory and Methods Section.

Conle, F. A. and Chu, C-C. Fatigue analysis and the local stress–strain approach in complex vehicular structures. *International Journal of Fatigue*, **19**: 317–323, 1997.

Cremona, C. Reliability updating of welded joints damaged by fatigue. *International Journal of Fatigue*, **18**: 567–575, 1996.

Doerk, O., Fricke, W. and Weissenborn, C. Comparison of different calculation methods for structural stresses at welded joints. *International Journal of Fatigue*, **25**: 359–369, 2003.

Fermér, M. and Svensson, H. Industrial experiences of FE-based fatigue life predictions of welded automotive structures. *Fatigue and Fracture of Engineering Materials and Structures*, **24**: 489–500, 2001.

Jono, M. Fatigue damage and crack growth under variable amplitude loading and counting method of stress ranges. In *The International Conference on Cumulative Fatigue, Seville*, 27–29 May 2003.

Juditsky, A., Hjalmarsson, H., Benveniste, A., Delyonand, B., Ljung, L., Sjöberg, J. and Zhang, Q. Nonlinear black-box models in system identification: mathematical foundations. *Automatica*, **12**: 1725–1750, 1995.

Katajamäki, K., Lehtonen, M., Marquis, G. and Mikkola, T. Fatigue stress FEA round robin: Soft toe gusset on I-beam flange. In *Design and Analysis of Welded High Strength Steel Structures*, pp. 69–96, Engineering Materials Advisory Service, Stockholm, 2002.

Kozin, F. and Bogdanoff, J. L. Probabilistic models of fatigue crack growth: Results and speculations. *Nuclear Engineering and Design*, **115**: 143–171, 1989.

LuValle, M. J. and Copeland, L. R. A strategy for extrapolation in accelerated testing. *Bell Laboratories Technical Journal*, **3**: 139–147, 1998.

Maddox, S. J. and Razmjoo, G. R. Interim fatigue design recommendations for fillet welded joints under complex loading. *Failure and Fracture of Engineering Materials and Structures*, **24**: 329–337, 2001.

McEvily, A. J., Endo, M. and Murakami, Y. On the $\sqrt{\text{area}}$ relationship and the short fatigue crack threshold. *Fatigue and Fracture of Engineering Materials and Structures*, **26**: 269–278, 2003.

Miller, K. A historical perspective of the important parameters of metal fatigue; and problems for next century. In *Proceedings of the Seventh International Fatigue Congress*, Xue-Ren Wu and Zhong-Guang Wang (eds), pp. 15–39, Beijing, 1999.

NRIM. *NRIM Fatigue Data Sheets*. National Research Institute for Metals, Tokyo, 1979.

Paris, P. C., Tada, H. and Donald, J. K. Service load fatigue damage – a historical perspective. *International Journal of Fatigue*, **21**: S35–S46, 1999.

Pers, B.-E., Kuoppa, J. and Holm, D. Round robin test for finite element calculations. In *Welded High Strength Steel Structures*, A. F. Blom (ed.), pp. 241–250, Engineering Materials Advisory Service, London, 1997.

Schijve, J. Fatigue predictions and scatter. *Fatigue and Fracture of Engineering Materials and Structures*, **17**: 381–396, 1994.

Schijve, J. Fatigue of structures and materials in the 20th century and the state of the art. *International Journal of Fatigue*, **25**: 679–702, 2003.

Svensson, T. Prediction uncertainties at variable amplitude fatigue. *International Journal of Fatigue*, **19**: S295–S302, 1997.

Taylor, D. A mechanistic approach to critical-distance methods in notch fatigue. *Fatigue and Fracture of Engineering Materials and Structures*, **24**: 215–224, 2001.

Virkler, D. A., Hillberry, B. M. and Goel, P. K. The statistical nature of fatigue crack propagation. *Journal of Engineering Materials and Technology*, **101**: 148–153, 1979.

Wang, G. S. A probabilistic damage accumulation solution based on crack closure model. *International Journal of Fatigue*, **21**: 531–547, 1999.

10

Choice of Complexity in Constitutive Modelling of Fatigue Mechanisms

Erland Johnson and Thomas Svensson

10.1 Background

The product development process within the engineering and vehicle industry typically involves the following activities:

1. develop an idea for a product;
2. establish a specification of requirements for the product;
3. choose concepts for the different components of the product;
4. decompose the requirements to the component level;
5. establish design loads on the different components;
6. develop a design proposal for each of the components;
7. predict the properties of the different components;
8. optimize the properties of the different components;
9. develop and manufacture a prototype;
10. verify product properties and specification of requirements.

Following the fourth step, a concept with requirements is established for each component in the product. In the following four steps (steps 5–8) the purpose is to virtually develop an optimum product, i.e. without producing a physical prototype (performed in step 9 above). The optimality is defined in each individual case from concurrent requirements on both cost (e.g. material consumption) and performance (e.g. structural strength or endurance).

Robust Design Methodology for Reliability: Exploring the Effects of Variation and Uncertainty
edited by B. Bergman, J. de Maré, S. Lorén, T. Svensson
© 2009, John Wiley & Sons, Ltd

Today, this virtual optimization is commonly performed with the finite element method (FEM). A finite element program essentially needs three types of input data, namely:

1. loads;
2. geometry (design proposal);
3. material data (to be used in one of the predefined material models in the FE program or, alternatively, in separate user-defined material models that are called from the program).

The FE program uses this information to calculate the stresses and the displacements in the model. These values are then combined with fatigue data for the material in order to determine the fatigue life of the product. In practice the process is iterative where the engineer, when cost/performance requirements are not fulfilled, has to go back and perform modifications in geometric design, manufacturing technique, type of heat treatment or choice of material. Sometimes the engineer may enter an impasse and be forced to also reduce the requirements on, for instance, the load level. Modifications in the load or the geometry are connected to the input types 1 and 2 above while the remaining modifications are related to type 3, material data. The latter category involves changes on two different levels. When only modifications of the material are performed, for instance modifications in the manufacturing or heat treatment procedures, the material model could often be kept, while only the material parameters are changed. On the other hand, when the material is replaced, a change of material model might be necessary. Examples of material models are elasticity (with different subclasses regarding nonlinearity and anisotropy), plasticity (with different subclasses regarding hardening, e.g. isotropic or kinematic hardening) or viscoelasticity. Examples of material parameters are Young's modulus, Poisson's ratio, modulus of hardening and relaxation time. Some of these parameters are present in more than one material model (e.g. Young's modulus), while some material parameters are specifically connected to a certain material model (e.g. modulus of hardening or relaxation time).

A critical and recurrent issue in the iterative procedure described is to judge whether a certain design fulfils the requirements on performance and endurance or not. On what foundation is this decision taken? A number is obtained from the calculation program but in practice an uncertainty (sometimes large) prevails as to how to interpret the result. A consequence is that decisions are postponed quite frequently until a prototype has been manufactured. This is cost-demanding for several reasons:

1. Resources (computer equipment and man power) have been put into calculation activities, which are still not sufficient for decision making.
2. The product development time increases, which in itself corresponds to increased costs and also lost market time since the product cycles today are becoming shorter and shorter.
3. Increased costs for manufacturing several prototypes since physical prototypes to a larger extent become involved in the iterative process described above.

Parallel to the increased need for decisions based on virtual simulations, computers are continuously becoming even more powerful. The computer capacity today enables a discretization of a geometry into a large number of finite elements with a very small discretization error. The computer capacity also implies that extremely advanced and complex material models could be used. Today computers could, in contrast to the situation only 10 years ago, perform simulations with advanced material models while maintaining moderate execution times. At the same time as discretization errors decrease and an increased model complexity has become possible to simulate, however, it is easy to lose the perspective and disregard the fact that other sources of error have not decreased to the same extent. These other sources of error have instead become the parameters controlling the accuracy in the results and therefore also become the bottleneck for reaching a rational decision. The sources of error can be described as follows:

1. There is an uncertainty in the load representativity. The loads are chosen from a more or less guessed typical customer, or perhaps worst customer. Irrespective of how the loading is chosen, there is an uncertainty as to how much the chosen load agrees with the actual service loads.
2. In spite of the use of computationally complex material models, critical model simplifications might be introduced. There is a risk that the model complexity is not focused on the most dominant effects of the performance requirements of interest.
3. There is an uncertainty in the material parameter values, which usually increases with increasing model complexity since it becomes too expensive (too many laboratory tests) to maintain the accuracy in each material parameter for a more complex model compared with a model with fewer parameters. An increased level of complexity also reduces the application range of the model with a poorer prediction capability outside this range as a result.

Today there is a large difference in complexity between the models that are used in industry and those models that are developed within academic research activities. The industrial models are phenomenological with few material parameters, which are calibrated against standard tests. As an example, for a steel material, a plasticity model with linear (kinematic) hardening is often used and its material parameters are calibrated against tensile tests and cyclic tests. For high cycle fatigue, the models are usually based on Wöhler curves, calibrated against high cycle fatigue tests on smooth specimens. In contrast to this, the research literature has plenty of much more advanced material models. This difference between industry and academia could partly be explained by the fact that more complex material models require an increased amount of testing to determine the (larger number of) material parameters. This counteracts the industrial requirement of a reduced number of tests.

Is the increase in available computer capacity used in the most efficient way today? It is not possible to give a definite answer to this question. However, one thing is clear. Discretization and material modelling can today be performed with such a high accuracy by computer, in contrast to the situation only a decade ago, that from a product development perspective other bottle necks within virtual simulations have emerged. (It should, however, be mentioned that the use of complex material models requires a certain number of interpretations, which still give rise to scattering in numerical results (cf. Bernauer and Brocks, 2002.) To achieve sufficient accuracy in the FE calculations is usually no problem today, what is problematic, however, is to specify the total uncertainty in the complete product development chain and, from this, choose a discretization and material model with appropriate complexity and accuracy. This specified uncertainty is different at different stages of the product development. During earlier stages, focus is on rapidity at the sacrifice of accuracy. It might, for instance, be sufficient to identify areas with high stress without obtaining accurate information about the specific stress levels, while in the later stages a high accuracy is required, for example for verifications of fatigue life, and thereby more computationally demanding models are acceptable. In all stages there is, however, a need, from a specific accuracy standpoint, to choose an optimum discretization and material model. The influence of discretization on accuracy is well known today and depends, in practice, only on the setting of the numerical parameters in the FE program. The accuracy in a material model is, however, a more complex issue since it is based on test results. These involve scattering from the material itself as well as from the test procedure.

The discussion points to the fact that when choosing a material model, it is necessary to simultaneously consider all sources of errors in all calculation stages within the product development chain, and, based on given resources for laboratory testing, identify the optimum material model.

10.2 Questions

The discussion above leads to the following questions:

1. How does the total uncertainty in the prediction of stress and fatigue life depend on the complexity (the number of parameters) in the chosen material model? Moreover, how complicated material models is it profitable to introduce in a certain situation?
2. Is it a cost effective industrial strategy to keep to the simpler models or could testing methodology be developed (towards more similarity with the service life situation) such that current testing costs in combination with more advanced material models could increase the total prediction accuracy? Is the application area of the model reduced below critical levels through this procedure?
3. Where should research efforts within virtual testing and simulations be directed in order to develop more efficient and better life predictions in the future?

10.3 Method

The different sources of error when predicting the fatigue life are illustrated in Figure 10.1. In order to understand the influence of the complexity of the material model on the total prediction inaccuracy, all sources of error must first be quantified. This cannot be done generally but a restriction to a component of a certain material, exposed to a certain load sequence, must be made. By choosing this as a typically occurring industrial case with an appropriate parameterization of the material behaviour, load sequence and geometry, it is judged that comparatively general conclusions could still be drawn. To obtain quantitative values of the different errors, different sources and methods, such as literature results, testing experience and calculation models, could be used. The total prediction error could then be expressed as a function of a variable number of parameters corresponding to a variable complexity in the material model. A general approach to the problem of complexity in empirical models is outlined in the next section.

10.4 Empirical Modelling

Modelling of physical phenomena are more or less always based on empirical observations. A mathematical formulation of such a model of a scalar measure of the phenomenon can be

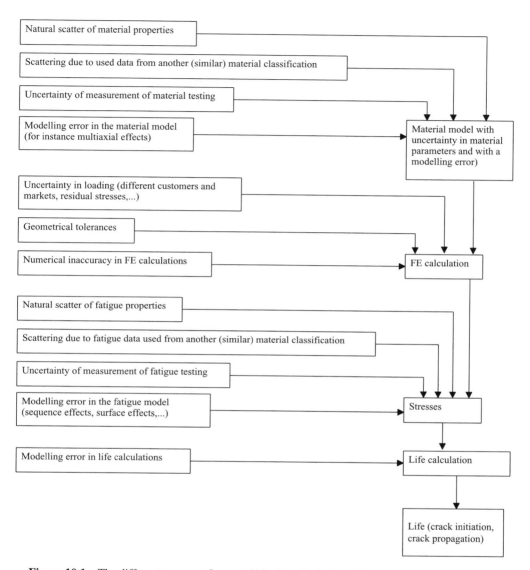

Figure 10.1 The different sources of error within the calculation of fatigue life for a component.

written

$$y = f(x_1, x_2, \ldots, x_m),$$

where $x_1, x_2, \ldots, x_m$ are different measurable variables that influence the phenomenon and f is an arbitrary mathematical formulation. By a Taylor expansion of the function around a nominal value $y_0 = \theta_0$ one can write

$$y \approx \theta_0 + \theta_1(x_1 - x_{1,0}) + \theta_2(x_2 - x_{2,0}) + \ldots + \theta_m(x_m - x_{m,0}) + \theta_{m+1}(x_1 - x_{1,0})^2 + \ldots$$
$$(10.1)$$

where $x_{.,0}$ are the nominal values of the influential variables, the parameters $\theta_1, \theta_2, \ldots$ are proportional to different derivatives of the original function f and the approximation can be done as good as we need to by adding terms of higher degree or limiting the domain of application. The series expansion of the function makes it somewhat easier to understand the problem of model complexity and we will use it here to demonstrate some fundamental problems in empirical modelling.

In the case of a pure empirical function, the partial derivatives are not known and the method for determining the function from observations is to estimate the parameters $\theta_0, \theta_1, \theta_2, \ldots$. This is usually done by means of some least square method, i.e. by minimizing the squared errors between the observed values and the model fit. Such a fit can be done arbitrarily close to the observations by choosing more variables, and in the limit one can obtain a perfect fit by choosing the same number of variables as the number of observations. However, in such a limiting case the modelling is quite useless, since no data reduction has been made. Further, no information is left for judging the uncertainty in the model and consequently nothing is known about the quality of future predictions based on the model. This fact gives rise to the complexity problem in modelling: What is the optimal trade-off between model complexity and prediction ability?

10.5 A Polynomial Example

The following example from simple polynomial regression on one variable demonstrates the complexity problem: We have observed ten values $y_1, y_2, \ldots, y_{10}$ depending on one variable x and want to find the function

$$y = f(x).$$

A Taylor expansion of the function is

$$y = \theta_0 + \theta_1(x - x_0) + \theta_2(x - x_0)^2 + \ldots, + \theta_{p-1}(x - x_0)^{p-1} + e, \qquad (10.2)$$

where e is the error in the model, which represents both neglected x terms and other unknown or neglected influences to the measure y. By assuming a random occurrence of such influences in the observations, one can model the error term e as a random variable. This is the general statistical approach, providing tools for estimating both confidence bands for the estimated parameters, and confidence- and prediction bands for the model.

The $n = 10$ observations of y obtained for ten reference values of x give the parameter estimates $\hat{\theta}_0, \hat{\theta}_1, \hat{\theta}_2, \ldots, \hat{\theta}_{p-1}$ and we now want to decide how many parameters one should use in the model. Of course, the more parameters are included the better fit one can obtain, but what about the possibilities for prediction? Using the statistical approach we can search for the number of parameters that gives the best prediction abilities. If the random variable is assumed to have a Gaussian distribution, the following prediction limits will contain 95% of future measurements $\tilde{y}$:

$$\tilde{y}(x) = \hat{y}(x) \pm t_{0.025,n-p} \cdot s \cdot \sqrt{1 + g_{n,p}(x; \mathbf{x}_{ref})}, \qquad (10.3)$$

where $\hat{y}(x)$ is the estimated value based on the model (10.2) and the estimated parameters $\hat{\theta}_0, \hat{\theta}_1, \hat{\theta}_2, \ldots, \hat{\theta}_{p-1}$, $t_{0.025,n-p}$ is the 2.5% quantile in the Student-t distribution with $n - p$ degrees of freedom, s is the estimated standard deviation of the random variable e, $g_{n,p}$ is a function of the value of the influential variable x for the actual prediction situation and the values of the reference vector $\mathbf{x}_{ref}$ used in the estimation procedure. The theory behind this formula and an expression for the g function can be found in ordinary text books on linear regression.

Figures 10.2 shows the results of using the model (Equation 10.2) with different polynomial degrees, corresponding to different number of parameters. Each figure shows the ten observations as dots, the fitted polynomial function as a line, and 95% prediction limits around the fitted function calculated using Equation (10.3).

In the polynomial of zero degree the estimated function is simply $y = \hat{\theta}_0$ and the given prediction band is expected to contain 95% of new observations whose influential variable is chosen within the interval $0 \leq x \leq 1$. The first degree polynomial fit unexpectedly gives a wider prediction band. This is the result of the fact that the reference measurements happened to give a very weak slope, and the improvement of the fit does not compensate enough for the more uncertain parameter estimates. The second degree polynomial does not show any improvement, but the third degree clearly shows a narrower prediction band and thereby a substantial improvement in prediction ability. The fourth degree polynomial gives a better fit to the observed values, but the prediction band now expands again, due to the uncertainties in the parameter estimates, and finally the fifth degree performs even worse by means of prediction purposes. Further increased complexity by choosing higher degrees is not shown here, but they give even wider prediction bands. One can conclude from the examples shown in Figure 10.2 that the best choice of model complexity seems to be the third-degree polynomial, i.e. the model with four parameters, as this model gives the best prediction ability.

This simple polynomial example demonstrates the strength of statistical methods when choosing an optimal complexity for physical empirical models. However, it can be done more rigorously and rationally by a formal criterion. The conclusion of optimal complexity in the example was based on a visual judgement about the area of the prediction bands in Figure 10.2. This area is a rough representation of the expected prediction variance and is an intuitive picture of one of the formal criteria of the optimal choice that has been presented in the literature, namely the Breiman–Friedman criterion (Breiman and Freedman, 1983), which minimizes the estimated expected prediction variance $\hat{U}^2_{n,p}$:

$$\hat{U}^2_{n,p} = s^2_{n,p}\left(1 + \frac{p}{n - p - 1}\right), \tag{10.4}$$

where $s_{n,p}$ is the estimated standard deviation for the random error in the empirical model, i.e. the error e in our polynomial example

$$s^2_{n,p} = \frac{1}{n - p}\sum_{i=1}^{n}(y_i - \hat{y}_i)^2,$$

where y_i is the ith observed value and $\hat{y}_i$ is the predicted value. This simple criterion can be shown to be precisely the expected prediction variance in the case of normally distributed

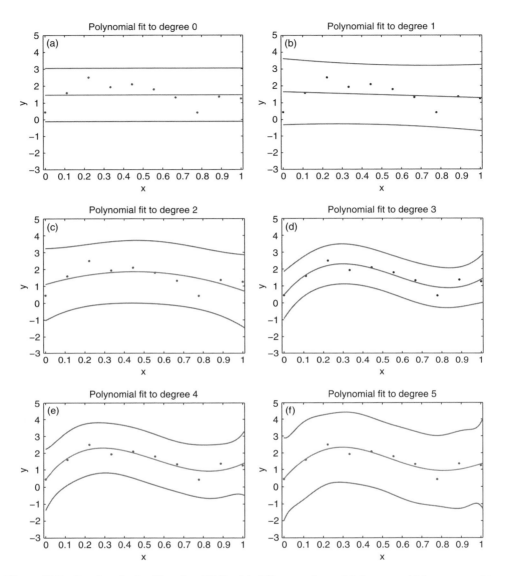

Figure 10.2 Results of using Equation (10.2) with different polynomial degrees, which correspond to a different number of parameters.

variables in the function y. In the polynomial example this is not fulfilled, since nonlinear transformations of a variable cannot be normally distributed if the variable itself is normal. However, the criterion may still be a good approximate rule for decision making, which will be seen in the example.

The estimated standard deviation $s_{n,p}$ depends greatly on the complexity, since an increase in the number of parameters p corresponds to smaller errors e, but it also depends on the number of reference observations n through its precision. On the other hand, the second term in parentheses increases with increasing complexity and thereby the criterion gives a trade-off between model complexity and prediction uncertainty.

In the example, the square roots of the prediction variances were estimated at:

p	1	2	3	4	5	6
$\hat{U}_{10,p}$	0.67	0.78	0.80	0.52	0.67	0.94

and the Breiman–Friedman criterion produces the same result as the visual inspection, i.e. the minimum is obtained for $p = 4$, the third-degree polynomial.

Here, only increasing degrees of complete polynomial models have been tested. Of course, in a real case, one should also try any subset of the different terms to find the optimal choice, but the limited choice still demonstrates the point: Optimal complexity can be found by means of statistical modelling of prediction uncertainty.

10.6 A General Linear Formulation

A generalization of the statistical idea behind the example gives the following model for a general empirical relationship:

$$Y = E[Y] + \sum_{i=1}^{p} \theta_i (X_i - E[X_i]) + \sum_{j=p+1}^{m} \theta_j (X_j - E[X_j]) + \underbrace{\underbrace{\sum_{k=m+1}^{\infty} \theta_k (X_k - E[X_k])}_{\tau}}_{e}$$

Here each influential variable is regarded as a random variable X with its expected value $E[X]$, which means that if the model is applied on the expected values for all X variables, then the result will be the expected value of Y, $E[Y]$. A certain variable X may be a function of some other variable, which gives the possibility to include also nonlinear terms into the model.

The sum is partitioned into three parts: the first one with p parameters is the model with optimal complexity, chosen by the criterion above; the second part contains the $m - p$ variables that are neglected in the model, because their influence is not large enough to add any useful information to the model; and the third part contains an unknown number of variables that are not known or not measurable and thereby represent the random contribution to the model.

A certain chosen model with p parameters can then be written:

$$Y = \theta_0 + \sum_{i=1}^{p} \theta_i (X_i - E[X_i]) + e,$$

where the error e may be modelled as a random variable if the model is used on a population of neglected influential variables $\{X_k; k = p + 1, p + 2, \ldots\}$. This means that if the outcome of the X variables are randomly chosen according to an appropriate distribution, then the statistical procedure with confidence and prediction intervals will be valid for each future use of the model on the same population.

The given abstract approach for treatment of empirical models gives an insight into complexity problems and is helpful for the judgement of the validity of a certain model. However, in a specific situation, the assumptions behind the approach are difficult to fulfil. The difficulties are as follows:

1. The behaviour may violate the assumptions about Gaussian distributions of the X variables.
2. True representative populations of the X variables are not easy to establish, which makes the random choice of reference tests difficult. This is in particular true for the unknown influential variables $\{X_k; k = m + 1, m + 2, \ldots\}$.
3. The values of X variables included in the model may not be known completely in reference tests, but are subject to uncertainty.

These difficulties must be considered in each specific situation, and necessary approximate solutions decided. The problem of nonlinearities and non-Gaussian distributions of the variables can in a specific case be overcome by replacing the simple criterion (Equation 10.4) with simulated prediction variances.

10.7 A Fatigue Example

We will here apply the given approach to the problem of modelling high cycle fatigue life. In order to diminish nonlinearity problems, we choose a log transformation and regard the logarithm of the fatigue life as a linear function of a number of influential variables:

The Basquin model $\ln N = \theta_0 + \theta_1 \ln \Delta S$

Neglected $\qquad + \theta_2 f_2(S_m) + \theta_3 f_3(seq) + \theta_4 f_4(T) + \theta_5 f_5(freq)$

Not measurable $\quad + \begin{cases} \theta_6 f_6(a_0) + \theta_1 f_1(C) + \theta_8 f_8(S_{op}) + \theta_9 f_9(G_{size}) \\ + \theta_{10} f_{10}(G_{conf}) + \theta_{11} f_{11}(G_{orient}) + \theta_{12} f_{12}(HV_{local}) \\ + \sum_{k=13}^{\infty} \theta_k(X_k - E[X_k]) \end{cases}$

In this formula the first line is the logarithm of the classic Basquin equation: $N = \alpha \cdot \Delta S^{-\beta}$ with $\theta_0 = \ln \alpha$ and $\theta_1 = -\beta$. The second line contains the influential variables mean stress, load sequence, temperature and loading frequency. The following two lines show examples of known influential variables which are usually not measurable in the design stage, namely the initial crack length, the crack geometry, local grain properties such as size, configuration and orientation, and local hardness. Finally, the sum in the last line represents all influential variables that are not known in the fatigue damage process. This choice of model complexity, neglecting all influential variables except the load range, is usual in industrial applications and in Chapter 7 this is roughly motivated by complexity arguments.

The presented formulation gives a special interpretation of systematic model errors in simplistic empirical modelling. When using the simple Basquin model one will avoid systematic model errors if the reference tests for the estimation of $\{\theta_1, \theta_2\}$ are performed on a random choice of neglected variable values, i.e. using for instance mean load values from the population of those load mean values that will appear in future applications of the model. If one succeeds in making such a choice, the reference test will provide information about the variability around the model and, with additional distribution assumptions, confidence and prediction limits can be calculated.

The resulting variability may be too large for an efficient design and then improvements of the model must be considered. This can be done in two different ways:

1. Increase the complexity in the model by including some of the neglected variables. This will *decrease the number* of variables in the error term e and thereby give possibilities for more precise predictions, but it demands more reference tests and more measurements of variables.
2. Narrow the application of the model to a subset of the population and place corresponding restrictions on the future use of the model. This will *decrease the variance* of the existing variables in the error term e and give more precise predictions.

A typical combination of these two solutions is to use different Basquin equations for different classes of stress mean values, i.e. partition the population of mean values into a small number of subsets and estimate one pair of Basquin parameters $\{\theta_1, \theta_2\}$ for each subset. This will result in larger overall complexity, since more parameters are used, without the need for estimating any parameter for the mean value influence.

References

Bernauer, G. and Brocks, W. Micro-mechanical modelling of ductile damage and tearing- results of European numerical round robin. *Fatigue and Fracture of Engineering Materials and Structures*, **25**: 363–384, 2002.

Breiman, L. and Freedman, D. How many variables should be entered in a regression equation? *Journal of the American Statistical Association*, **78** (381): 131–136, 1983. Theory and Methods Section.

11

Interpretation of Dispersion Effects in a Robust Design Context

Martin Arvidsson, Ida Gremyr and Bo Bergman

11.1 Introduction

In this chapter we discuss an interpretation of dispersion effects in unreplicated two-level fractional factorials. We propose that dispersion effects may be seen as interactions between control factors and uncontrolled and unobserved factors. This interpretation clarifies the value of dispersion analysis in robust design experimentation. Here a design is considered to be robust if it has a smaller deviation from a specified target value than other designs considered. Thus, it is recognized that, in the use of a product or process, the output will not be exactly at the specified target value but will deviate from it. Note that in general the target value might be a function of what is usually called a signal factor. For different values of the signal factors we want the output to be as close to the corresponding target value as possible.

Besides signal factors, two other categories of factors affecting the performance of a product or process are important for understanding the concept of robust designs. In the first category we have control factors, which are controlled and specified by a designer. The second category comprises what are referred to as noise factors, which are factors that are difficult, impossible and/or very expensive to control in real life. A robust design may be achieved by means of robust design experimentation in which interactions between control factors and noise factors are identified and exploited in order to minimize the deviation from the target value. For further discussion of this issue readers are referred to Box et al. (1988), Bergman and Holmqvist (1989), Shoemaker et al. (1991) and Box and Jones (1992).

It is not possible in an experiment to include or observe all noise factors, either because of their nature or simply due to the fact that they are unknown. It is argued in this chapter that

---

This chapter is based on the article 'Interpretation of Dispersion Effects in a Robust Design Context,' *Journal of Applied Statistics*, Vol. 33, 2006, published by Routledge, Taylor & Francis Group.

Robust Design Methodology for Reliability: Exploring the Effects of Variation and Uncertainty
edited by B. Bergman, J. de Maré, S. Lorén, T. Svensson

interactions between control factors and uncontrolled and unobserved factors are manifested as dispersion effects, an interpretation that has consequences for the identification of dispersion effects. This interpretation of dispersion effects has been indicated in Abraham and MacKay (1992), Freeny and Nair (1992), Hynén (1994), Engel and Huele (1996) and Steinberg and Bursztyn (1998).

11.2 Dispersion Effects

To create robust designs it is important to have knowledge of noise factors, and in particular their interactions with control factors. The main concern in this chapter is noise factors that cannot be controlled or observed in an experimental situation.

11.2.1 Factor Characteristics

The ability to gain knowledge about interactions between noise factors and control factors depends on certain characteristics of the noise factors; consider Figure 11.1 as a starting point for a discussion of these characteristics.

In Figure 11.1, the square indicates the set of all conceivable factors. The right-hand circle represents all factors that are controllable in some sense. Some of these factors, called *controllable noise factors*, are possible to control only in experimental conditions. The remainder of the controllable factors are denoted *control factors*; the levels of these factors are set at a certain level by a designer, thus making them controllable during operating conditions.

The left-hand circle in Figure 11.1 represents the factors that influence the response under study. Besides control factors and controllable noise factors, there is basically one other type of influential factor and that is noise factors that cannot be controlled even under experimental conditions due to their nature or due to the fact that they are unknown to the experimenter. This third type of factor will for the sake of brevity be denoted *random noise factors*.

Figure 11.1 shows factors present in an experimental situation as well as in operating conditions; in addition there are factors that are present only in an experimental situation. Thus it is fruitful to make a division between operating conditions and experimental conditions. Certain factors associated with the experimental situation but not present under operating conditions are better held constant during an experiment. In contrast, variation in random

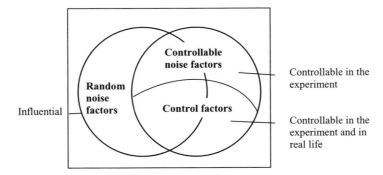

Figure 11.1 Categories of factors affecting a product (based on León et al., 1993).

noise factors present during operating conditions should be allowed so that the influence of these factors can be analysed, as argued for example by Taguchi (1986).

11.2.2 Dispersion Effects as Interactions

Interactions between control factors and noise factors are of great interest in robust design experimentation as they can be used to diminish the effect of noise factors on product output. To exemplify, let us first assume a very simple model for the output, y, with only one control factor, C, and one controllable noise factor, N:

$$y = \beta_0 + \beta_1 C + \alpha N + \gamma C N = \beta_0 + \beta_1 C + \eta. \tag{11.1}$$

In Equation (11.1), η is the total effect of N on result y, i.e. $\eta = (\alpha + \gamma C)N$. Thus, $V(\eta) \propto (\alpha + \gamma C)^2$ is minimized by choosing $C = -\alpha/\gamma$. However, in some cases, the model is unknown to the experimenter, which is typically the case if the difference in response variance is caused by a random noise factor R instead of N. If the levels of R can be observed by the experimenter, its corresponding effect on y can be studied by the use of regression analysis (see Freeny and Nair, 1992). If the levels of R cannot be observed by the experimenter, the variance of η for different values of C should be estimated instead. The difference in variance between the high and low levels of C for the simple example in this section is:

$$V(C = 1) - V(C = -1) = (\alpha + \gamma)^2 - (\alpha - \gamma)^2 = 4\alpha\gamma. \tag{11.2}$$

From Equation (11.2) it can be understood that a dispersion effect of C will be identified if C interacts with a random noise factor with an influential effect.

To further explore the nature of dispersion effects, imagine a model for the output y with two control factors, C_1 and C_2, and two random noise factors, R_1 and R_2:

$$
\begin{aligned}
y = {} & \beta_0 + \beta_1 C_1 + \beta_2 C_2 + \beta_{12} C_1 C_2 + \alpha_1 R_1 + \alpha_2 R_2 \\
& + \gamma_{11} C_1 R_1 + \gamma_{12} C_1 R_2 + \gamma_{21} C_2 R_1 + \gamma_{22} C_2 R_2 + \alpha_{12} R_1 R_2 \\
= {} & \beta_0 + \beta_1 C_1 + \beta_2 C_2 + \beta_{12} C_1 C_2 + \eta.
\end{aligned} \tag{11.3}
$$

In Equation (11.3) $\eta = (\alpha_1 + \gamma_{11} C_1 + \gamma_{21} C_2)R_1 + (\alpha_2 + \gamma_{12} C_1 + \gamma_{22} C_2)R_2 + \alpha_{12} R_1 R_2$ is the effect of R_1 and R_2 on the output y. Thus,

$$V(\eta) = (\alpha_1 + \gamma_{11} C_1 + \gamma_{21} C_2)^2 + (\alpha_2 + \gamma_{12} C_1 + \gamma_{22} C_2)^2 + \alpha_{12}^2 V(R_1 R_2). \tag{11.4}$$

Analogous to the operation in Equation (11.2) above, we now look at the difference in this variance when C_i is at a high level and a low level:

$$V(C_i = 1) - V(C_i = -1) = 4\alpha_1 \gamma_{i1} + 4\alpha_2 \gamma_{i2}, \quad C_i, i = 1, 2 \tag{11.5}$$

From Equation (11.5) we conclude that a dispersion effect can be identified if for example C_1 interacts with R_1 and R_1 also has an influential effect, and/or if C_1 interacts with R_2 and that factor is influential.

In the general case with l random noise factors and k control factors, η can be written as:

$$\eta = \sum_{j=1}^{l} \left(\alpha_j + \sum_{i=1}^{k} \gamma_{ij} C_i \right) R_j + \sum_{m} \sum_{m \neq n} \alpha_{mn} R_m R_n. \tag{11.6}$$

Further the variance of η is

$$V(\eta) = \sum_{j=1}^{l} \left(\alpha_j + \sum_{i=1}^{k} \gamma_{ij} C_i \right)^2 + V\left(\sum_{m} \sum_{m \neq n} \alpha_{mn} R_m R_n \right). \tag{11.7}$$

Finally the difference in variance between a high and a low level of a control factor C_i, $i = 1, \ldots, k$ is:

$$V(C_i = 1) - V(C_i = -1) = 4 \sum_{j=1}^{l} \alpha_j \gamma_{ij}. \tag{11.8}$$

Thus, a dispersion effect may be viewed as a sum of interaction effects between a control factor and random noise factors with influential effects.

11.2.3 Identification of Dispersion Effects

The problem of identifying dispersion effects from unreplicated fractional factorials was originally studied by Box and Meyer (1986). Since then a large number of researchers have suggested alternatives and improvements to the Box–Meyer method: see Bergman and Hynén (1997), Brenneman and Nair (2001) and McGrath and Lin (2001a). Lately Brenneman and Nair (2001) and Schoen (2004) have shown that methods available for dispersion effect identification are sensitive to assumptions made on the underlying variance model. If dispersion effects are interpreted as argued in this chapter, the method suggested by Brenneman and Nair (2001) for experiments with additively modelled variance seems to be the most suitable.

11.3 Discussion

This chapter argues that dispersion effects can be viewed as manifested interactions between control factors and random noise factors with main effects. The usefulness of identifying control factors with an influence on the dispersion is twofold. First, by using the fact that the dispersion is smaller for certain levels of the control factors, a more robust design can be achieved. Second, an identified dispersion effect carries valuable information. For an experimenter, this dispersion effect can be seen as an indication of an active random noise factor, and resources can be directed towards identifying this factor. If this previously unknown factor can be identified and controlled in an experiment, it can be included in a second round of experiments as a controllable noise factor. In this way more precise information about this factor and its effect on the response can be obtained.

The assumption of a correctly identified location model is a common denominator in all methods for identification of dispersion effects in unreplicated fractional factorials. If this assumption is inaccurate, aliasing between location and dispersion effects may complicate not only the identification of dispersion effects but also the interpretation of dispersion effects. In fact, McGrath and Lin (2001b) showed that two unidentified location effects create a spurious dispersion effect in their interaction column. McGrath (2003) proposed two procedures for determining whether a dispersion effect is real or whether it actually stems from two unidentified location effects. Moreover, Bisgaard and Pinho (2003–2004) discussed possible follow-up strategies to investigate whether identified dispersion effects are actually due to aberrant observations or faulty execution of some of the experimental runs. Use of the procedures suggested by Bisgaard and Pinho (2003–2004) and McGrath (2003) is advisable before dispersion effects are deemed real and interpreted as interactions between control factors and random noise factors. An interesting area for future research is to extend the interpretation of dispersion effects presented in this chapter to split-plot experiments, which are useful in robust design experimentation. Split-plot experiments make it possible to efficiently estimate control factors by noise factor interactions. An interesting aspect of the interpretation of dispersion effects suggested here is that it will affect the variance structure traditionally assumed for split-plot experiments.

References

Abraham, B. and MacKay, J. Discussion to Nair, V. N., (1992), 'Taguchi's parameter design: A panel discussion'. *Technometrics*, **34**(2): 127–161, 1992.

Bergman, B. and Holmqvist, L. A Swedish programme on robust design and Taguchi methods. In *Taguchi Methods – Proceedings of the 1988 European Conference*, Bendell, T. (ed.) Elsevier Science Publishers, Barking, 1989.

Bergman, B. and Hynén, A. Dispersion effects from unreplicated designs in the 2^{k-p} series. *Technometrics*, **39**(2): 191–198, 1997.

Bisgaard, S. and Pinho, A. Follow-up experiments to verify dispersion effects: Taguchi's welding experiment. *Quality Engineering*, **16**(2): 335–343, 2003–2004.

Box, G. E. P. and Jones, S. Split-plot designs for robust experimentation. *Journal of Applied Statistics*, **19**(1): 3–25, 1992.

Box, G. E. P. and Meyer, R. D. Dispersion effects from fractional designs. *Technometrics*, **28**(1): 19–27, 1986.

Box, G. E. P., Bisgaard, S. and Fung, C. An explanation and critique of Taguchi's contributions to quality engineering. *Quality and Reliability Engineering International*, **4**: 123–131, 1988.

Brenneman, W. A. and Nair, V. N. Methods for identifying dispersion effects in unreplicated factorial experiments: A critical analysis and proposed strategies. *Technometrics*, **43**(4): 388–405, 2001.

Engel, J. and Huele, A. F. A generalized linear modeling approach to robust design. *Technometrics*, **38**: 365–373, 1996.

Freeny, A. E. and Nair, V. N. Robust parameter design with uncontrolled noise factors. *Statistica Sinica*, **2**: 313–334, 1992.

Hynén, A. *Robust Design Experimentation*. Division of Quality Technology, Linköping University, 1994.

León, R. V., Shoemaker, A. C. and Kackar, R. N. Discussion to Coleman, D. and Montgomery, D. (1993) A systematic approch to planning for a designed experiment. *Technometrics*, **35**(1): 21–24, 1993.

McGrath, R. N. Seperating location and dispersion effects in unreplicated fractional factorials. *Journal of Quality Technology*, **35**(3): 306–316, 2003.

McGrath, R. N. and Lin, D. K. J. Testing multiple dispersion effects in unreplicated fractional factorial designs. *Technometrics*, **43**: 406–414, 2001a.

McGrath, R. N. and Lin, D. K. L. Confounding of location and dispersion effects in inreplicated fractional factorials. *Journal of Quality Engineering*, **33**: 129–139, 2001b.

Schoen, E. Dispersion-effects detection after screening for location effects in unreplicated two-level experiments. *Journal of Statistical Planning and Inference*, **126**: 289–304, 2004.

Shoemaker, A. C., Tsui, K. L. and Wu, J. Economical experimentation methods for robust design. *Technometrics*, **33**(4): 415–427, 1991.

Steinberg, D. M. and Bursztyn, D. Noise factors, dispersion effects, and robust design. *Statistica Sinica*, **8**: 67–85, 1998.

Taguchi, G. *Introduction to Quality Engineering – Designing Quality into Products and Processes*. Asian Productivity Organization, Tokyo, 1986.

12

Fatigue Damage Uncertainty

Anders Bengtsson, Klas Bogsjö and Igor Rychlik

12.1 Introduction

Material fatigue is one of the most important safety issues for structures subject to cyclic loads and it is the cause of failure in a majority of cases. Fatigue is a two-phase process starting with the initiation of microscopic cracks in the material, and in the second phase these cracks continue to grow up to a critical size at which fracture occurs. In large structures cracks often exist from the beginning and in such a case the growth of these may be computed using fracture mechanics. However, in classic fatigue life computations, the material is assumed to be originally free of cracks. Fatigue life of a component is greatly influenced by a number of component- and material-dependent factors, such as geometry, size of the structure, surface smoothness, surface coating, residual tensions, material grain size and inner defects. Further, the nature of the load process is important. The complex dependence between these factors and fatigue life makes predictions uncertain and even for controlled laboratory experiments the results from fatigue life tests exhibit a considerable scatter, especially for high cycle life.

At the design stage there exist, depending on application, different strategies to design such that fatigue failure should not occur. When materials such as steel alloys or titanium, with a distinct fatigue limit, are used, the simplest approach is to design in such a way that the stresses are below, with some safety factor, the fatigue limit of infinite life. Other materials, for example aluminium, do not exhibit a clear threshold below which fatigue does not occur. Furthermore, recent research has shown that fatigue failure may occur for stresses below the fatigue limit when the number of cycles is very large (Bathias, 1999).

Another approach is to design the component for some finite life after which the component is replaced. Since fatigue exhibits a stochastic nature, the estimated life, expressed in number of load cycles or in time, milage, etc., must be connected to some measure of risk of failure, e.g. in the automotive industry when a component is designed according to a criterion that less than two failures out of 10^5 components should occur before 200,000 km of use. Another typical situation is assessment of safety of existing structures. Even here the criterion is often

Robust Design Methodology for Reliability: Exploring the Effects of Variation and Uncertainty
edited by B. Bergman, J. de Maré, S. Lorén, T. Svensson
© 2009, John Wiley & Sons, Ltd

formulated in terms of probabilities, or safety indexes (see Rychlik and Rydén, 2006) for a discussion, namely, one requires that the failure probability due to fatigue is for example 10^{-4}. To estimate such a risk, empirical models for time to failure are often used, such as Wöhler curves and the Palmgren–Miner's linear damage rule or the Paris–Erdogan crack growth equation (Paris and Erdogan, 1963). The methods give predictors of fatigue life and the simplest approach to assess the safety is to require that the predicted time to failure is longer than the design life. How much longer it should be depends on the safety level required and the different sources of uncertainties affecting the value of the predictor.

In this chapter we present a simplified safety analysis showing how the different sources of uncertainties can be combined into a safety index using a Bayesian approach with material- and component-dependent parameters modelled as random variables. To be able to estimate the expected damage and the covariance function of the damage rate from one sample path (measurement of a load), loads are assumed to be ergodic. We further concentrate on so-called symmetrical loads when the local narrow band approximation, proposed in Bogsjö and Rychlik (2009), can be used to estimate the expected damage. Both Gaussian and non-Gaussian loads will be considered. Note that for stationary random loads when the probabilistic model is known, one can obviously use Monte-Carlo methods to estimate the expected damage and variance and in such a case the ergodicity assumption is not necessary. In this chapter we will use probabilistic models to check the accuracy of our method. The proposed method is, however, more useful when only one (measured) load sequence is available.

We shall give most attention to the variability of the load relevant for fatigue. This is usually described by means of rainflow matrices, or rainflow range distributions, which measure the size of oscillation at a randomly chosen time during the design life. The distribution is then used to compute the average growth of fatigue damage. The computation of such distribution is a complicated task. For stationary Gaussian loads the rate of damage growth is constant and several approximations (see Bengtsson and Rychlik (2009) for a review), have been proposed to compute the expected rate of damage growth if the power spectral function is known.

For stationary Gaussian loads, when the power spectral density is uncertain and we deal with the crack growth problem (the slope in Wöhler curve is about -3), the damage growth depends on mean load value (here assumed to be zero), variance of the load and variance of its derivative (measuring the rate of loads oscillations). It is often assumed that, under stationary conditions, the uncertainties in prediction of fatigue life (related to load uncertainties) are sufficiently described if one knows the statistical errors of estimates of the damage growth rate. However, as demonstrated in Bengtsson and Rychlik (2009), prediction errors caused by variable material properties, geometry and modelling errors are not negligible in the computation of the safety index.

When studying a variable environment, the average damage growth rate alone may not be sufficient to properly estimate the risk for fatigue failure. For example, the risk of failure for a component of an oil platform during one year depends on the age of the component, and can be high during a year if an extreme storm hits the offshore structure. The probability of such a storm can be very small but may still influence the value of the estimated risk. Consequently the uncertainties in long term variability of load properties should be included in the risk analysis. In practical applications the uncertainties are measured by means of the variance of the accumulated fatigue damage. The computation of the variance is much harder than that of the mean. We shall give an example of such an analysis in connection with offshore loads.

The chapter is organized as follows. In Section 12.2 some basic definitions of rainflow damage are given and in Section 12.2.1 variable amplitude tests are discussed. Failure functions and a safety index are introduced in Section 12.3. In Section 12.4 estimation of expected damage and variance of damage are discussed: Nonparametric estimation methods are treated in Section 12.4.1, an approximation using a simpler damage rate is given in Section 12.4.2 and damage due to a variable environment is discussed in Section 12.4.3 and illustrated with an ocean load example. In the last section we give two examples of non-Gaussian loads: The first is measured stresses with a bimodal spectrum from a vessel sailing under stationary conditions for 30 minutes. It will be demonstrated that this measured stress is slightly non-Gaussian, which considerably affects the coefficient of variation of accumulated damage. The results will also be compared with a Gaussian load and a skewed load having the same spectrum. In the second example we consider vehicle fatigue damage caused by rough roads where the road profile is modelled as a Gaussian process with superimposed random irregularities.

12.2 Fatigue Review

Fatigue testing of components traditionally has been carried out using constant amplitude stress cycles. In these experiments, the stress oscillates between a minimum and maximum value until fatigue failure occurs. The shape of the cycles is usually sinusoidal. However, in most cases the shape has a small influence on the result. Repeating the experiments for different amplitudes, keeping the ratio, R, between minimum and maximum load constant, results in what is known as a Wöhler curve, also called the S–N curve, with a log-linear dependence between the number of cycles to failure, N, and the stress cycle range, s,

$$\log(N) = a - k \log(s) + e, \tag{12.1}$$

where parameters $a > 0$ and $k \geq 1$ depend on material and component properties and the stress ratio R. Properties which have little or no influence in the static case, such as the smoothness of the metal surface, residual tensions and size and geometry of the structure, may greatly influence the fatigue life. The total error due to scatter in material and component properties, modelling errors, etc., is represented by the term $e \in N(0, \sigma_e^2)$.

In real applications, however, loads are often not constant amplitude sinusoidals but instead are random processes, such as ocean wave loads or vehicle loads due to a random road surface. Here we assume that the material stress is proportional to the external load with a constant c which depends on the structure, and often approximative methods, such as the Finite Element Method (FEM), are needed for the computation of the constant.

For random processes the load cycles and cycle ranges need to be defined using some cycle count procedure. As with the constant amplitude load only local maxima and minima are of interest, the shape of the cycles has been shown to be less important. A number of cycle count procedures exist such as the 'crest-to-trough' count, which defines a cycle as a pair comprising local maximum and the preceding local minimum, or the 'positive peaks' count, where cycles are defined for each positive local maximum with cycle range equal to twice the value of the local maxima. However, in fatigue analysis, the 'rainflow' method, defined below, has been shown to give the most accurate results. The method was introduced originally by Endo and published in Japanese by Matsuishi and Endo (1968). Here we shall use the alternative definition given in Rychlik (1987), which is more suitable for statistical analysis.

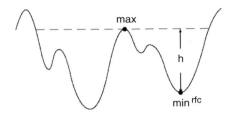

Figure 12.1   A rainflow pair

In the rainflow cycle count, each local maximum of the load process is paired with one particular local minimum, determined as follows:

1. From the local maximum one determines the lowest values in forward and backward directions between the time point of the local maximum and the nearest points at which the load exceeds the value of the local maximum (Figure 12.1).
2. The higher of those two minimum values is the rainflow minimum paired with that specific local maximum, i.e. the rainflow minimum is that which falls the least before reaching the value of the local maximum again on either side.
3. The cycle range, h, is the difference between the local maximum and the paired rainflow minimum.

Note that for some local maxima, the corresponding rainflow minimum could lie outside the measured load sequence. In such situations, the incomplete rainflow cycle constitutes the so-called residual and has to be handled separately. In this approach, we assume that in the residual the maxima form cycles with the preceding minima.

Fatigue damage from variable amplitude (random) loads is commonly regarded as a cumulative process. Using the linear Palmgren–Miner damage accumulation rule (Palmgren, 1924; Miner, 1945) together with Equation (12.1), we can define the total damage $D^{tot}(t)$ at time t as

$$D^{tot}(t) = \sum_i e^{-a}\, (c\, h_i)^k. \tag{12.2}$$

Parameters $a > 0$ and $k \geq 1$ are the same parameters as in Equation (12.1) and have traditionally been estimated using constant amplitude tests for some value of the stress ratio R. Since the ratio R for the cycles of a random process varies, the estimation of parameters a, k is difficult and in the literature different correction factors have been proposed. Fatigue failure is normally predicted when $D^{tot}(T) = 1$, although in practice fatigue failure may occur for $D^{tot}(T)$ between, say, 0.3 and 3. Hence, a direct application of Equation (12.2) may give less accurate predictions. Other factors to consider are, as listed in Johannesson et al. (2005a), that

- the order of the load cycles is important
- a constant amplitude load imposes residual stresses in the material to a higher extent than the variable amplitude load

- threshold stress levels, i.e. the stress levels necessary to cause crack growth, are assumed to be different for a random amplitude load.

A possible solution to incorporate these factors in the model is to estimate the parameters a, k of the S–N (Stress, Number of cycles to failure) curve using tests with variable amplitude loads similar to the real load processes. We will discuss this issue further in the next section.

12.2.1 Variable amplitude S–N curve

Let us introduce the equivalent cycle range defined as

$$h_n^{eq} = \left(\frac{1}{n} \sum_{i=1}^{n} h_i^k \right)^{1/k}, \tag{12.3}$$

where $\{h_i\}_1^n$ are load cycle ranges obtained by rainflow filtering, i.e. small cycles, with a range smaller than some chosen threshold relative to the fatigue limit, are removed and n is the number of remaining rainflow cycles in the blocked test load.

Empirical tests, see Agerskov (2000) and Figure 12.2, have shown that the S–N relation (Equation 12.1) is valid also for Gaussian random loads if the constant stress range s is replaced by $s^{eq} = ch_{n_0}^{eq}$,

$$\log(N) = a - k \log \left(c \cdot h_{n_0}^{eq} \right) + e, \tag{12.4}$$

where n_0 is the number of rainflow cycles in the blocked test load. Since parameters a, k are unknown, they have to be estimated ($h_{n_0}^{eq}$ depends on k, which makes the estimation more difficult, see Johannesson et al., 2005a, 2005b), and are thus uncertain. Using Bayesian ideas, we let parameters a and k be random variables, asymptotically normally distributed. (The variance of the error, σ_e^2, is also estimated from data and uncertain but we neglect this fact.) The variable amplitude experiments in Agerskov (2000) were performed for two different types of loads: real measurements of stress histories in different steel structures and simulated loads from Gaussian spectra using a Markov Chain of Turning Points (MCTP) technique. For loads consisting of repeated load blocks containing n_0 rainflow cycles, such as with the first type of experiments using real load histories, the equivalent cycle range is an uncertain value too. If n_0 is sufficiently large, such that

$$h_{n_0}^{eq} \approx \lim_{n \to \infty} h_n^{eq} = h^{eq}, \tag{12.5}$$

the block effect in testing can be neglected. Theoretically, simulated loads used in S–N experiments are ergodic and have relatively short memory, and hence $h_{n_0}^{eq}$ converges relatively fast to h^{eq}. However, in variable amplitude experiments in Johannesson et al. (2005b) there was some rainflow filtering performed on the signal and the load was approximated by a finite sum of random cosine processes and consequently the estimate $h_{n_0}^{eq}$ used in regression differs from the theoretical value h^{eq}. In addition we do not know if a new sequence was used for each experiment. Such issues may affect the regression analysis. However, here we assume that this can be neglected.

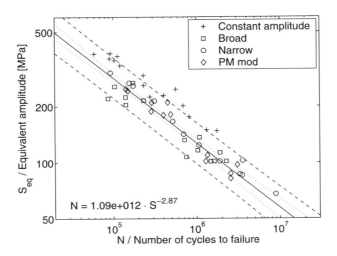

Figure 12.2 S–N curve estimated from variable amplitude tests using broad-banded, narrow-banded and Pierson–Moskowitz spectra compared with constant amplitude tests (Agerskov, 2000).

12.3 Probability for Fatigue Failure – Safety Index

Assume that the fatigue failure is predicted to occur before time T if the accumulated damage $D^{tot}(T) > 1$. In practice $D^{tot}(T)$ is replaced by the estimate of the expected damage $E[D^{tot}(T)]$. This can be rephrased by saying that load $Y(t), t \in [0, T]$, is safe for fatigue if $E[D^{tot}(T)] < 1$. In some situations one wishes to measure the risk for failure more accurately and to check whether the probability for failure of a component before time T, say one year, is below 10^{-4} or some other small value. To measure small risks one needs to consider different sources of uncertainties, e.g. uncertainties in geometry, material, statistical errors in estimation of the S–N curve as well as modelling errors. In the following, we sketch a simplified approach to combining different sources of uncertainties into one reliability measure.

When computing the safety index and coefficient of variation, it is more convenient to use the so-called *pseudo damage* $D(T) = e^a c^{-k} D^{tot}(T)$. Further, the stress proportional constant, c, is an uncertain parameter as a result of variability in geometry, manufacturing variability between components and model error. For simplicity of computations we assume that c is log-normally distributed with known expectation and variation coefficient.

The risk of fatigue failure thus depends on the following parameters or random variables: a, k (material strength), c (stress proportionality − geometrical uncertainties), $D(T)$ accumulated pseudo damage, e material variability (residuals between S–N data and fitted S–N curve). A function, $G(a, k, c, D(T), e)$, of these variables is called a failure function if

$$\begin{cases} G < 0 : \text{fatigue failure} \\ G = 0 : \text{limit state} \\ G > 0 : \text{nonfailure region.} \end{cases} \qquad (12.6)$$

Suppose now we are interested in the risk of fatigue failure in the time period $[0, T]$. If we then use the S–N relation, we can define the failure function as

$$G(a, k, c, D(T), e) = a - k \log(c) + e - \log(D(T)), \tag{12.7}$$

see Bengtsson and Rychlik (2009) for a more detailed presentation. Note that parameters a and k are uncertain (statistical errors) since those are estimates and usually based on a small number of tests, and e represents the residuals when fitting the S–N curve.

A simple measure of reliability against fatigue failure is the so-called Cornell's safety index, measuring the 'distance' from the mean location to failure expressed in number of standard deviations and defined by

$$I_C = \frac{E[G]}{\sqrt{V[G]}}. \tag{12.8}$$

Since the failure function is not uniquely defined, Cornell's safety index depends on how $G(a, k, c, D(t), e)$ is chosen. We will neglect this problem.

Still, computation of $V[G]$ is a complicated problem and we therefore propose to use the Gauss approximation to estimate its values, which leads to

$$
\begin{aligned}
I_C &= \frac{E[a - k \log(c) + e - \log(D(T))]}{\sqrt{V[a - k \log(c) + e - \log(D(T))]}} \\[2mm]
&\approx \frac{E[a] - E[k] \log(E[c]) - \log(E[D(T)])}{\sqrt{V[a] + \log(E[c])^2 V[k] + \dfrac{E[k]^2}{E[c]^2} V[c] + V[e] + \dfrac{V[D(T)]}{E[D(T)]^2} + 2C(k, D(T)) \dfrac{\log(E[c])}{E[D(T)]}}}.
\end{aligned} \tag{12.9}
$$

In the last row, we use the Gauss approximation and the assumption that the random variables $a, k, c, e, D(T)$ are independent except for k and $D(T)$. (In practice, the way regression is performed, a and k are correlated. For the sake of simplicity of presentation we neglect this here.) For the first four variables, mean and variances are derived by means of regression and residual analysis. The variability of a, k and e were studied in Johannesson et al. (2005a). Here we shall focus on the estimation of the variance and expectation of damage, $V[D(T)]$ and $E[D(T)]$.

In our approach, k is a random variable describing statistical uncertainties due to the finite (usually small) number of fatigue tests as well as some scatter of material properties. The load is also modelled as a random process since it is not known exactly in advance. The pseudo damage depends both on load and parameter k, i.e. it depends on k even if the load and exponent k are statistically independent. Consequently a way to compute $E[D(T)]$ and $V[D(T)]$ is to condition on k, namely:

$$E[D(T)] = E[\, E[D(T)|k]\,], \qquad V[D(T)] = V[\, E[D(T)|k]\,] + E[\, V[D(T)|k]\,],$$

where $E[D(T)|k]$, $V[D(T)|k]$ are functions of k, equal to the mean and the variance of $D(T)$ when k has a known fixed value. Similarly, conditioning on k, we can compute the covariance $C(k, D(T)) = E[kE[D(T)|k]] - E[k]\, E[E[D(T)|k]]$.

12.4 Computation of $\mathbf{E}\,[D(T)|k]$ and $\mathbf{V}\,[D(T)|k]$

When a model for a random load is fully specified, for example if the load is a stationary Gaussian process with a known mean and spectrum, then both expected damage as well as the coefficient of variation can be estimated by means of the *Monte Carlo* method. More precisely, one may simulate a large number of loads, with fixed length, from the model and compute the accumulated damage for each load. Then expected damage and variance can be estimated using standard statistical estimators. (Even the distribution of the damage can be studied in this way.) For large T, the observed damage is approximately normally distributed and consequently the mean and coefficient of variation fully describe the variability. However, the convergence to normality can be very slow in T, especially in the case of high values of parameter k.

In the situation when one has only one measured load, one could, based on data, fit a model and use Monte Carlo methodology to estimate mean and variance. However, the possibility of modelling error should not be neglected. A nonparametric approach, where no particular model (except ergodicity) for the measured load is specified, will be presented next.

12.4.1 Nonparametric Estimate

Suppose that the load is stationary and ergodic. (Ergodicity means that one infinitely long realization defines the random mechanism that generated the load.) In this section we shall use the fact that the damage $D(t)$ is absolutely continuous, i.e. the derivative $\dot{D}(t)$, called the damage rate, exists and $D(T) = \int_0^T \dot{D}(t)dt$. Further, suppose that the damage rate is a stationary process too. (This is not true for the rainflow damage for t close to the beginning of the measured load. However we neglect this fact here.)

The expected stationary damage rate $d = \mathrm{E}\left[\dot{D}(t)|k\right]$ is referred to as the damage intensity and the expected damage $\mathrm{E}[D(T)|k] = Td$. If the load has been observed up to time t, then a natural estimate of d, for a given k, is

$$\hat{d} = \frac{D(t)}{t}. \tag{12.10}$$

We turn now to the estimate of the variance of damage where we shall use an alternative formula for the rainflow damage employing the damage accumulation rate $\dot{D}(t)$. Let $Y(t)$ be the load process. As shown in Rychlik (1993a), one can write the rainflow damage rate as follows

$$\dot{D}(t) = k(Y(t) - Y^-(t))^{k-1}\dot{Y}(t), \tag{12.11}$$

where $Y^-(t)$ is the lowest value in the backward direction between t and the nearest exceedance of $Y(t)$ – see Figure 12.3.

Suppose that the process $X(t) = \dot{D}(t)$ is stationary and ergodic and let $r(s) = \mathrm{C}(X(0), X(s)|k)$ be the covariance function of X given k, then

$$V\left[\int_0^T X(t)\,dt|k\right] = \int_0^T \int_0^T \mathrm{C}(X(s), X(t)|k)\,ds\,dt = T\int_0^T 2\left(1 - \frac{s}{T}\right)r(s)\,ds = T\,\sigma^2. \tag{12.12}$$

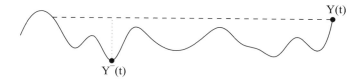

Figure 12.3 Definition of $Y^-(t)$.

Obviously σ^2 is a function of T as well. If $r(s)$ converges fast to zero as s increases, then this dependence can be neglected for large values of T and $\sigma^2 \approx 2\int_0^\infty r(s)ds$. However, in practice, $r(s)$ is unknown and has to be estimated from only one, often short, measurement of load.

Suppose the damage rate $X(s)$, $0 \leq s \leq t \leq T$, has been observed for a time period $[0, t]$ and let $\hat{r}(s)$ be an estimate of the covariance function $r(s)$. For a fixed s the estimate $\hat{r}(s)$ can be very uncertain, unless s is much shorter than t. In addition, since we assumed that $r(s)$ converges fast to zero, then a large part of $\hat{r}(s)$, say above t_0, is not significantly different from zero. Consequently we propose to remove the nonsignificant $\hat{r}(s)$ from the integral and estimate σ^2 by the following formula

$$\hat{\sigma}^2 = 2\int_0^{t_0}\left(1 - \frac{s}{t}\right)\hat{r}(s)\,ds. \tag{12.13}$$

Note that since $\hat{d}$ and $\hat{\sigma}$ are estimated from measured loads, they are uncertain and this has an influence on the safety index. One solution to this problem could be to compute an approximative confidence interval for the coefficient of variation. We will, however, not do this here.

12.4.2 Holm–de Maré Damage Rate

At the design stage, one is often interested in fast estimates of the expected damage and the coefficient of variation. This can be achieved by approximating the rainflow damage rate by replacing $Y(t) - Y^-(t)$ in Equation (12.11) by $|2Y(t)|$, namely

$$\dot{D}^{HM}(t) = k\,|2\,Y(t)|^{k-1}\,\dot{Y}(t)^+, \qquad x^+ = \max(0, x). \tag{12.14}$$

The damage defined by Equation (12.14) is known as Holm–de Maré damage since it was first introduced in Holm and de Maré (1988).

It can be shown (Rychlik, 1993b) that for any load, the accumulated damage $D(T) \leq D^{HM}(T)$. Except for the zero-mean symmetrical load, i.e. a load having symmetrical unimodal crossing intensity, the Holm–de Maré is often a very conservative bound and hence seldom used. Furthermore, in the case of stationary symmetrical loads, $E[D^{HM}(T)]$ coincides with the so-called narrow band damage introduced by Bendat (1964).

For symmetrical loads, however, estimating the fraction σ/d using the Holm–de Maré damage rate (Equation 12.14) gives approximately the same result as with the rainflow damage rate (Equation 12.11) (see Bengtsson et al., 2009). For asymmetrical loads the approximation needs to be investigated further. In Section 12.5.1 we shall give an example with a stationary load having skewness 0.8 and kurtosis 5.

12.4.3 Gaussian Loads in a Variable Environment

Consider a zero mean Gaussian load, $Y(t)$, and let $h_s(t) = 4\sqrt{V[Y(t)]}$ be the significant range at time t, while

$$f_z(t) = \frac{1}{2\pi}\sqrt{\frac{V[\dot{Y}(t)]}{V[Y(t)]}}$$

is the apparent frequency (the intensity of zero level upcrossings by Y). Since the significant range and mean period vary in time as the environment changes, it is convenient to introduce the normalized load $Z(s)$, which has $h_s(s) \approx 1$ and $f_z(s) \approx 1$. The normalized load $Z(s)$ is defined as

$$Z(s) = \frac{Y(g^{-1}(s))}{h_s(g^{-1}(s))}, \quad g(t) = \int_0^t f_z(x)\,dx \tag{12.15}$$

and, obviously, $Y(t) = h_s(t)Z(g(t))$. The normalized and scaled process $Z(s)$ has variance $V[Z(s)] = 1/16$ and variance of derivative $V\left[\dot{Z}(s)\right] = (\pi/2)^2$.

For stationary loads, the damage intensities and variances of damage for Y and Z are related as follows. Let $d, \tilde{d}$ be damage intensities (rainflow, Holm–de Maré) caused by load Y and the normalized load Z, respectively. Furthermore, let $\sigma^2, \tilde{\sigma}^2$ be variances defined in Equation (12.12), for the load Y, Z, respectively, then

$$d = f_z\,h_s^k\,\tilde{d}, \quad \sigma^2 = h_s^{2k}\,f_z\tilde{\sigma}^2. \tag{12.16}$$

Studying historical data, one often observes, however, that values of $h_s(t)$ and $f_z(t)$ change in a rather unpredictable manner, and then one may also model the variability of $h_s(t)$, $f_z(t)$ by means of random processes. Although the properties (parameters) of a random process describing the load Z may depend on h_s, f_z, we neglect this for the sake of simplicity and assume that the normalized random load Z is stationary and independent of the random processes h_s, f_z. This is often the case in offshore engineering, for example, when one is assuming a sea spectrum of Pierson–Moskowitz, or Joint North Sea Wave Observation Project (JONSWAP), type. In such a particular case, Z can be taken as a stationary process with a standardized Pierson–Moskowitz spectrum independent of the $h_s(t)$, $f_z(t)$ processes.

Loads can often be well modelled as zero mean stationary Gaussian processes for quite a long period of time t, say, when the environment, e.g. storm strength, changes, giving another value of significant range h_s and apparent frequency f_z, i.e. the damage rate changes. Suppose the environment changes slowly (relatively to f_z), such that the load can be regarded as 'locally' stationary, i.e. $h_s(t)$, $f_z(t)$ can be considered approximately constant for some (short) period of time. Further, assume short memory in the sense that the covariance $r(s)$ converges fast, relatively to length of the local stationarity, to zero. Consequently – as it was motivated in Bengtsson and Rychlik (2009), where the normalized process Z defined by Equation (12.15) was employed – the variance of the damage can be written as

$$V[D(T)] \approx \int_0^T \int_0^T C(h_s^k(t)f_z(t)\,\tilde{d}, h_s^k(s)f_z(s)\,\tilde{d})\,dt\,ds + \int_0^T E[h_s^{2k}(t)f_z(t)\tilde{\sigma}^2]\,dt, \tag{12.17}$$

while the expected damage can be approximated by

$$E[D(T)] \approx \int_0^T E[h_s(t)^k f_z(t) \tilde{d}] \, dt, \tag{12.18}$$

where $\tilde{d}$ is the damage intensity and $\tilde{\sigma}^2$ the variance of damage for the stationary process Z. For stationary and Gaussian Z, the upper bound of $\tilde{d}$ is $2^{-k/2}\Gamma(1 + k/2)$, equal to the damage intensity using the Holm–de Maré damage rate (Rychlik, 1993b). Note that the bound is independent of Z.

Finally, conditioning on k, the term $V[D(T)]/E[D(T)]^2]$ in the safety index (Equation 12.9) can be estimated by means of

$$\frac{V[D(T) \mid k]}{E[D(T) \mid k]^2} \approx \frac{\int_0^T \int_0^T C\left(h_s^k(t) f_z(t), \, h_s^k(s) f_z(s) \mid k\right) dt \, ds}{(\int_0^T E[h_s^k(t) f_z(t) \mid k] \, dt)^2} + \frac{\tilde{\sigma}^2}{\tilde{d}^2} \frac{\int_0^T E[h_s^{2k}(t) f_z(t) \mid k] \, dt}{(\int_0^T E[h_s^k(t) f_z(t) \mid k] \, dt)^2}, \tag{12.19}$$

where the first term represents the uncertainty of the environment and the second term describes uncertainties in fatigue accumulation during stationary periods.

As mentioned in Section 12.4.2, for symmetrical loads Holm–de Maré damage agrees with the narrow band approximation, which postulates (for stationary Gaussian loads) that the cycle amplitudes are Rayleigh distributed (see Rychlik, 1993b). If all cycles had equal duration $(1/f_z)$ and cycle amplitudes were independent, then $\tilde{\sigma}/\tilde{d}$ would be equal to

$$\sqrt{\frac{\Gamma(1 + k)}{\Gamma(1 + k/2)^2} - 1}, \tag{12.20}$$

which, for $k = 3, 4, 5$, takes the values 1.55, 2.24, 3.14, respectively. Employing Equation (12.20), only variability of $f_z(t)$ and $h_s(t)$ needs to be modelled in order to compute the mean and coefficient of variation of the accumulated damage. For narrow band spectra, however, some wave grouping may occur and in such a case the assumption of independent cycle amplitudes makes the approximation of Equation (12.20) less accurate.

12.4.4 Example

In the following we shall compute Equations (12.18) and (12.19) using the model for $h_s(t)$ proposed in Baxevani et al. (2005). Consider a fixed position in the North Atlantic Ocean. It has been found that in the Northern Atlantic

$$\log(h_s(t)) = a_0 + a_1 \sin(\phi t) + a_2 \cos(\phi t) + e(t) = a_0 + m(t) + e(t), \quad \phi = \frac{2\pi}{365.2}, \tag{12.21}$$

with t in days, and where $e(t)$ is a zero mean Gaussian process with constant variance σ_H^2.

For a fully developed sea the Pierson–Moskowitz spectrum is often used. Then one has that $2.5\sqrt{h_s} < f_z^{-1} < 3.5\sqrt{h_s}$, and hence a simplification of $h_s^k(t) f_z(t)$ is proposed by taking $f_z^{-1} = 3\sqrt{h_s}$. Obviously, this is a crude approximation and more elaborate studies are needed to propose a better model. In model (12.21) we have

$$E[h_s(t)^\beta \mid \beta] = e^{\beta m(t)} e^{a_0 \beta + \beta^2 \sigma_H^2/2}. \tag{12.22}$$

(Here we have used that for $X \in N(m, \sigma^2)$, $E[e^{\beta X} | \beta] = e^{\beta m + \beta^2 \sigma^2/2}$.) Then for a fixed T and $\beta > 0$

$$\int_0^T E\left[h_s^\beta(t) \mid \beta\right] dt = e^{a_0\beta + \beta^2\sigma_H^2/2} \int_0^T e^{\beta m(\tau)} d\tau = T J_0(a\beta \cdot i)e^{a_0\beta + \beta^2\sigma_H^2/2},$$

where J_0 is the Bessel function of the first kind while $a = \sqrt{a_1^2 + a_2^2}$. Based on satellite observations (see Baxevani et al., 2005), $(a_0, a, \sigma_H^2) = (0.87, 0.39, 0.16)$ at 53°W and 50°N. Thus at this location, with T equal to one year measured in seconds and taking a typical value $k = 3$, we have

$$\int_0^T E\left[\frac{1}{3}h_s^{k-1/2}(t) \mid k\right] dt = \frac{T}{3} \cdot e^{2.675} J_0(0.975 \cdot i) = 1.91 \cdot 10^8,$$

$$\int_0^T E\left[\frac{1}{3}h_s^{2k-1/2}(t) \mid k\right] dt = \frac{T}{3} \cdot e^{7.205} J_0(2.145 \cdot i) = 3.6 \cdot 10^{10}. \tag{12.23}$$

Consequently the second term in Equation (12.19) is equal $1.0 \cdot 10^{-6}(\tilde{\sigma}/\tilde{d})^2$ which is clearly negligible. However, the term increases with k and for $k = 5$ it is $1.8 \cdot 10^{-5}(\tilde{\sigma}/\tilde{d})^2$ while for $k = 7$ the term is $1.1 \cdot 10^{-3}(\tilde{\sigma}/\tilde{d})^2$ which is no longe negligible. For example, using Equation (12.20) the second term in Equation (12.19) has the value 0.04 for $k = 7$ which is quite large.

To compute the first term in Equation (12.19) we need the correlation function for $\log(h_s(t))$. A typical $r(t)$, based on observations from US NODC Buoy 46005, that will be used in this example is

$$C\left(\log(h_s(t_2)), \log(h_s(t_1))\right) = \sigma_H^2\, e^{-0.0009(t_2-t_1)^2 - 0.009|t_2-t_1|} = r(t_2 - t_1), \tag{12.24}$$

where t_i are in hours. Now for any β

$$C\left(h_s(t_2)^\beta, h_s(t_1)^\beta \mid \beta\right) = e^{2\beta a_0 + \beta^2\sigma_H^2} e^{\beta(m(t_2)+m(t_1))} \cdot \left(e^{\beta^2 r(t_2-t_1)} - 1\right),$$

here t_i are in hours. Consequently, with $\beta = k - 1/2$, the first component in Equation (12.19) is equal to

$$3600^2 \frac{\int_0^T \int_0^T C\left(h_s(t)^\beta, h_s(s)^\beta \mid \beta\right) dt\, ds}{(J_0(a\beta \cdot i)e^{a_0\beta + \beta^2\sigma_H^2/2})^2}.$$

Now, for the considered location and time period T, equal to one year, the values of formula (12.19) for $k = 2, 3, 4, 5, 7$ are 0.0027, 0.012, 0.041, 0.16, 4.2, respectively. Only for $k = 7$ is the second term in Equation (12.19), having value 0.04, not negligible.

12.5 Non Gaussian Loads – Examples

In the previous section we considered Gaussian (and close to Gaussian) wave loads. We will now provide two examples where we estimate the coefficient of variation σ/d for loads that are non-Gaussian. First, damage from a measured stress, slightly non-Gaussian, is analysed and compared with a Gaussian load model and an asymmetrical load having the same spectrum as the measured stress, and then there is an example of fatigue damage to a vehicle due to the roughness of the road, using a road model consisting of a stationary Gaussian process with randomly added nonstationary irregularities.

12.5.1 Measured Stress in a Vessel

We shall analyse the uncertainty in the estimate of accumulated damage for a measured stress at some location in a vessel. The load is a superposition of wave-induced stresses and transients due to whippings. The estimated load power spectrum has two peaks, as can be seen in Figure 12.4 (upper left plot). From the same figure it also follows that the load is slightly

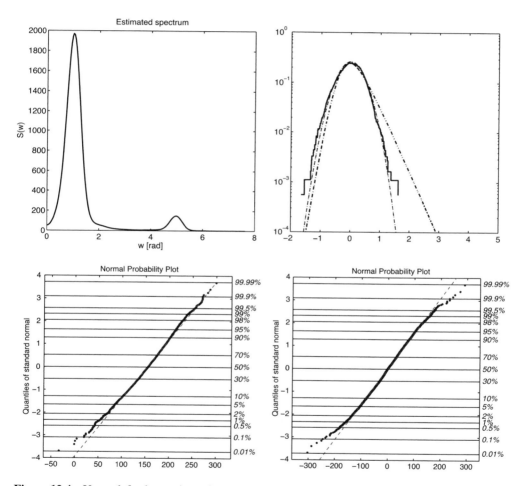

Figure 12.4 Upper left plot: estimated spectrum for the measured sea load. Upper right plot: the observed upcrossing intensity (irregular line), compared with theoretical upcrossing intensities for a Gaussian process with the estimated spectrum (dotted lines). Lower left plot: measured load plotted on normal probability paper. Lower right plot: estimates of the derivative of the load plotted on normal probability paper.

non-Gaussian (lower plots) but with a symmetrical crossing spectrum (upper right plot). Note that the observed upcrossing intensity is very close to the crossing intensity of a stationary Gaussian process having mean zero and with the same power spectrum as the observed load.

We will now investigate how well one can model the load by means of a Gaussian process. From the discussion above it seems reasonable that the observed nominal rainflow damage is close to the expected rainflow damage for the Gaussian load. This is also the case, and for $k = 3$ the observed damage is equal to $7.7 \cdot 10^8$ while the expected damage computed using the Monte Carlo method is $7.1 \cdot 10^8$ for the Gaussian load. We conclude that the damage in the observed load grows on average like a Gaussian load.

From the lower right plot in Figure 12.4, where the estimated $((Y(t + \Delta t) - Y(t))/\Delta t)$ derivatives for the measured load are plotted on normal probability paper, it can be seen that the derivative has larger tails than the Gaussian load. Although the measured load is only slightly non-Gaussian, the coefficient of variation for the measured load differs significantly, as will be shown later, from that of the Gaussian load, i.e. the damage accumulates on average as in a Gaussian load but varies more around the mean.

Thus, an interesting question arises about the influence of non-Gaussianity on the coefficient of variation. Here only one example will be studied which is inspired by Gao and Moan (2007), who investigated the fatigue damage for non-Gaussian asymmetrical loads with a bimodal spectrum. (One of the loads analysed in Gao and Moan (2007) had skewness 0.8 and kurtosis above 4.) In order to check how the non-Gaussianity may affect the coefficient of variation, we have simulated loads with the bimodal spectrum given in Figure 12.4, having skewness 0.8 and kurtosis 5. An algorithm to simulate non-Gaussian loads with a given spectrum, skewness and kurtosis is presented in Åberg et al. (2009). Two problems will be studied next. First how the skewness affects the coefficient of variation and then if the Holm–de Maré damage (Equation 12.14) can be used to compute the coefficient of variation for a truly asymmetrical (skewed) load.

12.5.1.1 Comparison of the Coefficient of Variation for the Observed Load and the Gaussian Model

First we have estimated the coefficient of variation for the observed load. This is done by means of formulas in Section 12.4.1 for the rainflow damage for the normalized and scaled load Z defined in Equation (12.15). The estimates of $\tilde{\sigma}/\tilde{d}$ for $k = 3, 4, 5$ are given in the first row of Table 12.1.

Table 12.1 Table of estimates for $\tilde{\sigma}/\tilde{d}$ using the Holm–de Maré damage. The two last rows are estimated using 1000 simulations of the Gaussian load and the skewed non-Gaussian load having the observed spectrum. Numbers in parentheses are the 'true' values of the coefficient of variation.

Description	$k = 3$	$k = 4$	$k = 5$
Measured 'sea load'	2.4	3.4	4.8
Gaussian bimodal	1.3 (1.3)	1.9 (1.8)	2.6 (2.5)
Non-Gaussian load	2.4 (2.2)	3.9 (3.6)	5.7 (6.1)

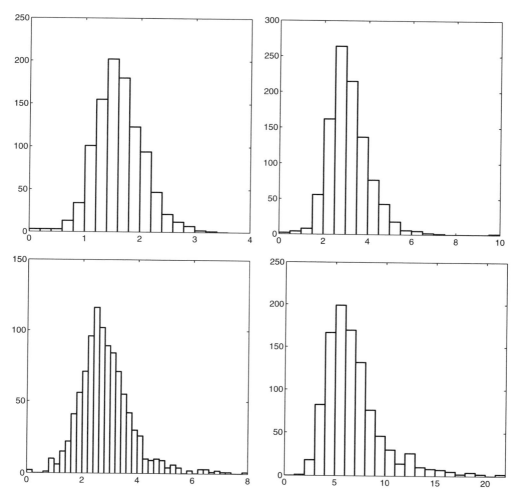

Figure 12.5 Upper plots: histograms of 1000 estimates of $\tilde{\sigma}/\tilde{d}$, $k = 3$ and $k = 5$, for a Gaussian load with spectrum from Figure 12.4. Lower histograms are computed for the non-Gaussian load with skewness 0.8 and kurtosis 5.

Next, the same procedure was repeated for 1000 simulations of a Gaussian process with the spectrum from Figure 12.4. The average estimates of $\tilde{\sigma}/\tilde{d}$ from 1000 simulations are given in the second row in the table. (For both loads the cut-off threshold t_0 in Equation (12.13) is chosen to be 120 while t is 1800 seconds.)

Clearly the estimates of $\tilde{\sigma}/\tilde{d}$ for the measured signal are higher than the values for the Gaussian load. In order to check whether the difference is significant and not just caused by statistical errors, due to the fact that only 30 minutes of the sea load has been measured, we present histograms of the 1000 estimates, using rainflow damage, computed for a Gaussian load (see upper row in Figure 12.5). From the histograms we conclude that the coefficient of variation of the measured load is significantly higher than that of the Gaussian load.

12.5.1.2 Coefficient of Variation for the Skewed Load

Let us first compare the crossing spectra for the three loads. In the upper right plot in Figure 12.4 we compare the observed number of upcrossings of the level u for the measured load (irregular line), the Gaussian load (smooth dashed line) and the skewed load (dashed irregular line). All signals have the same spectrum. The first observation is that all three functions have almost the same frequency of zero upcrossings f_z, i.e. the measured signal has $f_z = 0.245$ Hz while the Gaussian model and skewed model have $f_z = 0.263, 0.246$, respectively. The crossings of the Gaussian load and the measured signal are very close while the skewed load, although not very often (on average, once every three hours), reaches twice as high as the Gaussian load does. (Consequently for $k = 3$, the average damage of the skewed load is only 25% higher than the damage of the Gaussian load. This difference increases to 121% for $k = 5$.)

Further, as can be seen in Table 12.1, last row, the 'true' values of coefficients of variation (given in parentheses) are quite close to the estimates computed, using formulas in Section 12.4.1, for the Holm–de Maré damage. However, more investigations are needed to produce a general recommendation that the method can be used for any ergodic load with unimodal crossing intensity.

As previously advocated, the advantage of the proposed method, i.e. to estimate the coefficient of variation by means of Equation (12.10–12.13) and the Holm–de Maré damage, is that it can be applied even for quite short measurements of a load. (Obviously, if a very long signal is available, one could split it into shorter parts to compute mean and variance of the damage and hence the coefficient of variation.) However, if the measured load is very short, the estimates become uncertain. This is illustrated in Figure 12.5, where the histograms of the estimated coefficients of variation based on 30 minutes measurement are presented. We can see that the uncertainty is particularly large for $k = 5$.

12.5.2 Vehicle Fatigue Damage Caused by Rough Roads

In the example above with the measured sea load, which is a superposition of wave induced stresses and whippings (added transients), and hence slightly non-Gaussian, the observed variation coefficient was higher but of the same order as the Gaussian load. We will now give an example of a stationary load which has a much higher coefficient of variation than the loads studied in the previous section.

Stochastic models are commonly used to describe the randomness of measured road profiles. Vehicle models travelling on road profiles modelled as stationary Gaussian processes have been studied extensively (see, for example, Sun (2001) and Mucka (2004) for some recent studies). However, measured profiles are not accurately described by a stationary Gaussian model (Bogsjö, 2007). The road model used in this chapter includes random nonstationary irregularities.

Fatigue damage is assessed by studying a quarter-vehicle model travelling at constant velocity on road profiles (Figure 12.6). This very simple model cannot be expected to predict loads on a physical vehicle exactly, but it will highlight the most important road characteristics as far as fatigue damage accumulation is concerned. The parameters in the model are set to mimic heavy vehicle dynamics. The total force acting on the sprung mass is denoted $Y(t)$.

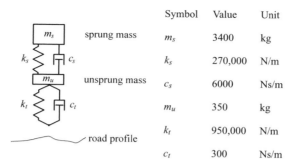

Symbol	Value	Unit
sprung mass	m_s	3400	kg
	k_s	270,000	N/m
	c_s	6000	Ns/m
unsprung mass	m_u	350	kg
	k_t	950,000	N/m
road profile	c_t	300	Ns/m

Figure 12.6   Quarter vehicle model.

An analysis of measured road tracks shows that actual roads contain short sections with above-average irregularity. Such irregularities are shown to cause most of the vehicle fatigue damage (Bogsjö, 2007). The main variability in the road is described by the stationary Gaussian process $Z_0(t)$, with spectrum

$$R_0(\xi) = \begin{cases} 10^{a_0} \left(\frac{\xi}{\xi_0} \right)^{-w_1}, & \xi \in [0.01, 0.20], \\ 10^{a_0} \left(\frac{\xi}{\xi_0} \right)^{-w_2}, & \xi \in [0.20, 10], \\ 0, & \text{otherwise.} \end{cases} \tag{12.25}$$

Irregularities of random shape, length and location are superimposed to $Z_0(x)$. The jth irregularity is denoted by $Z_j(x)$, $j > 0$, and the road with superimposed irregularities is denoted by $Z(x)$. To exemplify this, a 400 m long road is simulated and plotted in Figure 12.7.

The irregularities are modelled as (nonstationary) conditional Gaussian processes. To avoid discontinuities at the start and end of the rough sections, the added irregularities start and end with zero slope and zero level. The irregularities are simulated conditioning on the zero boundary values, see Bogsjö (2006) for more details. Furthermore, the location and length of the sections with added roughness are random. More precisely, the distance between the end of an irregularity and the start of the next is exponentially distributed. The irregularity length is also exponentially distributed.

Y is stationary but non-Gaussian. It can also be shown that it has a symmetrical crossing spectrum (see Figure 12.8), and consequently the Holm–de Maré damage can be used. In

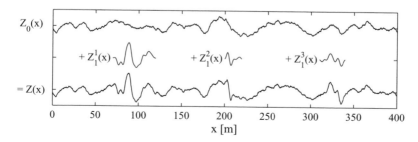

Figure 12.7   A synthetic (computer simulated) road profile $Z(x)$.

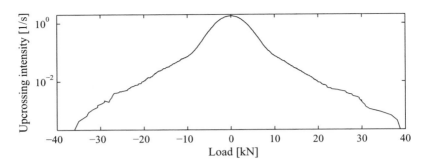

Figure 12.8 The observed upcrossing intensity in the load presented in Figure 12.10 (top plot).

Figure 12.9, the rainflow and the Holm–de Maré damage rates are compared. In this particular example, 400 m long roads are realized from the road model introduced in Bogsjö (2007), but here the irregularities are placed deterministically at the metre intervals [70, 120], [200, 220] and [310, 340]. The quarter vehicle is simulated when travelling at 70 km/h, so these intervals correspond to the following intervals expressed in seconds: [3.6, 6.2], [10.3, 11.3] and [15.9, 17.5]. Obviously, we expect that the damage rates are higher at these locations.

In the upper plot of Figure 12.9 both rates are computed for $k = 5$ from a particular load $Y(t)$, for $t \in [0, 20.6]$, and in the lower plot the average from 100 simulations is plotted. This illustrates that the two rates differ for a particular load, but on average they are quite similar.

In the following we will use the Holm–de Maré rate since its simplicity makes it advantageous. The upper plot in Figure 12.10 shows a road profile Z, simulated from the model described in Section 12.5.2. The middle plot shows the load $Y(t)$ acting on the sprung mass ($v = 80$ km/h). The load peaks correspond to road sections with increased roughness. The lower plot shows the Holm–de Maré damage intensity for $k = 3$.

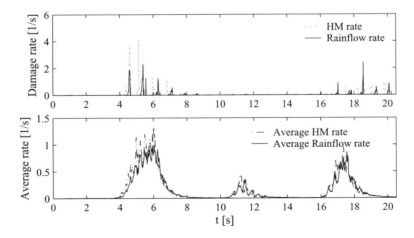

Figure 12.9 Comparison between the rainflow and Holm–de Maré (HM) damage rate, $k = 5$.

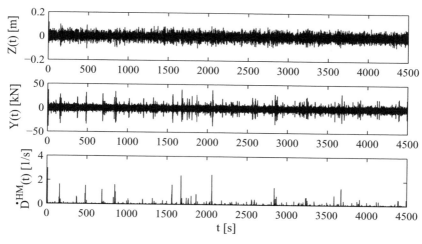

Figure 12.10 A road profile $Z(t)$, corresponding vehicle load $Y(t)(f_z = 1.79$ $[s^{-1}])$ when $v = 80$ km/h, and the damage intensity $\dot{D}(t)$ for $k = 3.\}$

The covariance function $r(t)$ can be estimated from a single time series $X(t)$ defined by Equation (12.14). In the example in Figure 12.11, $r(t)$ is estimated from the damage intensity in Figure 12.10. Then the variance and coefficient of variation (CoV) are estimated using formulas (12.10) and (12.13). A Monte Carlo study is now presented in order to show the accuracy of the CoV estimation. One hundred roads of length 100 km are realized with the expected distance of 400 m between irregularities and the expected length of an irregularity $\tau = 32$ m. The quarter vehicle is simulated travelling at 80 km/h. Then the empirical standard deviation and the average damage is computed using these 100 outcomes of $D(T)$. The empirical standard deviation divided by the average damage gives an estimate of the coefficient of variation of $D(T)$, which is referred to as the empirical coefficient of variation.

For comparison, we also compute from each load sequence the Holm–de Maré intensity and estimate its covariance function. Use of formulas from Section 12.4.1 gives us an estimate of the coefficient of variation from each load sequence. The average result is close to the empirical

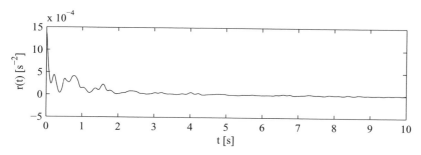

Figure 12.11 Covariance function, $r(t) = C(\dot{D}(t + t_0), \dot{D}(t_0)|k = 3)$ estimated from the sample path $\dot{D}(t)$ in Figure 12.10.

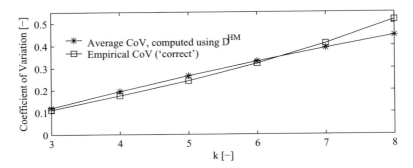

Figure 12.12   Average CoV computed using Equations (12.10) and (12.13), compared with the empirical CoV.

coefficient of variation, computed using the standard rainflow approach, as illustrated by Figure 12.12. Thus an average we will produce a coefficient of variation close to the true value. As expected, when k increases the coefficient of variation also increases. So for $k = 8$ the coefficient of variation is as high as 0.5, despite the length of the road ($L = 100$ km, $T = 1$ h 15 min). This has practical consequences; when vehicle loads have been measured on a 100 km long road, with these properties it is not enough to consider only the expected damage, since the variability is still quite high.

It is also interesting to study the uncertainty of the estimate. In Figure 12.13 a histogram shows all 100 estimates, compared with the empirical coefficient of variation. We conclude that the uncertainty of the estimated variation coefficient is quite large, even when the measurement is as long as 100 km.

Finally we compare the uncertainties in the estimates of the expected damage for the studied vehicle load with the wave loads presented in the previous section, see Table 12.1. The coefficient of variation of normalized and scaled load Z is 9.0, 19.9 and 29.3 for $k = 3, 5, 7$, respectively. As we can see this is about six times higher than for wave loads.

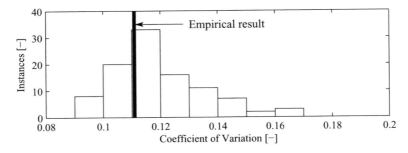

Figure 12.13   Uncertainty of CoV estimate, for $k = 3$.

References

Åberg, S. Podgórski, K. and Rychlik, I. *Fatigue Damage Assessment for a Spectral Model of Non-Gaussian Random Loads. Probabilistic Engineering Mechanics,* **24**: 608–617, 2009.

Agerskov, H. Fatigue in steel structures under random loading. *Journal of Constructional Steel Research,* **53**: 283–305, 2000.

Bathias, C. There is no infinite fatigue life in metallic materials. *Fatigue and Fracture of Engineering Materials and Structures,* **22**(7): 559–565, 1999.

Baxevani, A., Rychlik, I. and Wilsson, R. A new method for modelling the space variability of significant wave height. *Extremes,* **8**: 267–294, 2005.

Bendat, J. S. *Probability Functions for Random Responses: Prediction of Peaks, Fatigue Damage and Catastrophic Failures.* Technical Report, NASA, 1964.

Bengtsson, A. and Rychlik, I. Uncertainty in fatigue life prediction of structures subject to Gaussian loads. *Probabilistic Engineering Mechanics,* **24**: 224–235, 2009.

Bengtsson, A., Bogsjö, K. and Rychlik, I. Uncertainty of estimated rainflow damage for random loads. *Marine Structures,* **22**: 261–274, 2009.

Bogsjö, K. Development of analysis tools and stochastic models of road profiles regarding their influence on heavy vehicle fatigue. *Vehicle System Dynamics,* **44**(51): 780–790, 2006.

Bogsjö, K. *Road profile statistics relevant for vehicle fatigue.* PhD thesis, Department of Mathematical Statistics, Lund University, 2007.

Bogsjö, K. and Rychlik, I. *Vehicle Fatigue Damage Caused by Road Irregularities. Fatigue and Fracture of Engineering Materials and Structures,* **32**: 391–402, 2009.

Gao, Z. and Moan, T. Fatigue damage induced by non-Gaussian bimodal wave loading in mooring lines. *Applied Ocean Research,* **29**: 45–54, 2007.

Holm, S. and de Maré, J. A simple model for fatigue life. *IEEE Transactions on Reliability,* **37**: 314–322, 1988.

Johannesson, P., Svensson, T. and de Maré, J. Fatigue life prediction based on variable amplitude tests–methodology. *International Journal of Fatigue,* **27**: 954–965, 2005a.

Johannesson, P., Svensson, T. and de Maré, J. Fatigue life prediction based on variable amplitude tests–specific applications. *International Journal of Fatigue,* **27**: 966–973, 2005b.

Matsuishi, M. and Endo, T. Fatigue of metals subject to varying stress. In *Paper Presented to Japan Society of Mechanical Engineers,* Fukuoka, Japan, pp. 37–40. The Japan Society of Mechanical Engineers, March 1968. [In Japanese.]

Miner, M. A. Cumulative damage in fatigue. *Journal of Applied Mechanics,* **12**: A159–A164, 1945.

Mucka, P. Road waviness and the dynamic tyre force. *International Journal Vehicle Design,* **36** (2/3): 216–232, 2004.

Palmgren, A. Die Lebensdauer von Kugellagern. *Zeitschrift des Vereins Deutscher Ingenieure,* **68**: 339–341, 1924.

Paris, P. C. and Erdogan, F. A critical analysis of crack propagation laws. *Journal of Basic Engineering. Transactions of the ASME,* **D85**: 528–534, 1963.

Rychlik, I. Rain flow cycle distribution for fatigue life prediction under Gaussian load processes. *Fatigue and Fracture of Engineering Materials and Structures,* **10**: 251–260, 1987.

Rychlik, I. Note on cycle counts in irregular loads. *Fatigue and Fracture of Engineering Materials and Structures,* **16**: 377–390, 1993a.

Rychlik, I. On the 'narrow-band' approximation for expected fatigue damage. *Probabilistic Engineering Mechanics,* **8**: 1–4, 1993b.

Rychlik, I. and Rydén, J. *Probability and Risk Analysis : An Introduction for Engineers.* Springer-Verlag, Berlin, 2006.

Sun, L. Computer simulation and field measurement of dynamic pavement loading. *Mathematics and Computers in Simulation,* **56**: 297–313, 2001.

13

Widening the Perspectives

Bo Bergman and Jacques de Maré

13.1 Background

In the preceding chapters we have tried to widen the perspective on reliability improvement work with a special emphasis on variation as a source of unreliability. We have discussed efforts to make products and processes insensitive, or robust, against sources of variation and thereby enhancing their reliability and failure mode avoidance. We have also discussed how, by modelling uncertainty and variation, product dimensioning can be made and reliability enhanced. The important fatigue type failure modes have been especially emphasized. However, there are a number of other perspectives we have not yet discussed or only mentioned in passing.

First we note that, essentially, perspectives on reliability can be divided into two different groups, engineering perspectives and organizational perspectives. For the engineering perspectives we will, in addition to the ones already discussed earlier in this book, also discuss those related to the introduction of software in products and issues related to the maintaining of reliability of products in usage – usually discussed under headings such as 'Reliability Centred Maintenance' and 'Condition Monitoring'. Of course, there are still a large number of engineering perspectives on reliability that are not covered. We refer the reader to, for example, O'Connor (2002).

In the organizational perspectives it is emphasized that failures occur not only for technical or engineering reasons but also that the organizational contexts are important in order to understand the occurrence of failures. In Section 13.3 we will discuss consequences of system complexity and we will draw the attention to a new area, resilience engineering, and to the existence of what is sometimes called High Reliability Organizations (HRO). These organizations manage to operate in a very safe way, in spite of a very complex context with an abundance of possibilities of failures with serious consequences. Cultural aspects become vital.

Robust Design Methodology for Reliability: Exploring the Effects of Variation and Uncertainty
edited by B. Bergman, J. de Maré, S. Lorén, T. Svensson
© 2009, John Wiley & Sons, Ltd

In real life the two perspectives, engineering and cultural, do interact. In reliability, feedback of operational reliability experiences is important in order to enhance reliability growth. An important example is flight operations where a nonblaming culture has made failure reporting natural. Based on the feedback experiences, safety and reliability improvements have made passenger flights safe and reliable.

Another type of organizational perspective is how organizations adopt the ways of working we have suggested in this book. Surely, in some cases they have to adapt the ideas before they can adopt them. In fact, the adaptation process is an important and often necessary part in any successful adoption process. In Section 13.3 we will briefly discuss experiences of the industrialization of the ideas put forward in this book.

Finally, we will discuss our own acquisition of knowledge in the collabration, which has led to this book, and we will discuss some interesting future research objectives.

13.2 Additional Engineering Perspectives on Reliability

We have only scratched the surface when it comes to different engineering perspectives on reliability. This short section is by no means exhaustive. It draws attention only to some perspectives we find especially important and interesting from a robustness and variation point of view. One such interesting area is software reliability.

13.2.1 Software Reliability

Software failures are different from mechanical ones. For software, faults are created very early in the software development process, when the system engineer or, later, the programmer makes mistake that creates a fault ('bug'). A software fault is such that, when triggered by certain inputs, the software does not produce what is intended. It should be noted that not all faults lead to failures. This means that, first of all, it is important that the usage envelope of the product is well known at the very start of product development. This is not different from mechanical or electrical systems. If we want a product to have high reliability in usage, we need to understand the usage profile and all its sources of variation, especially noise factors. Also the next step of the development is similar to all other types of systems – creativity is an important aspect of the design of robust and reliable solutions. However, in software development it is hard to understand how something corresponding to parameter design (see Chapter 1) is created. Instead, the testing and verification step should be performed in a way that ensures that the system is functioning everywhere in the usage space. Of course, it is impossible to test all possible cases of the usage space.

An early attempt to cover the usage space has been via what has been called 'Statistical Usage Testing', which was originally suggested in connection with the so-called 'Cleanroom Methodology' for software development; see for example Runeson and Wohlin (1993). An alternative and an economically more feasible approach would be to utilize design of experiments to cover the multidimensional usage space in comparatively few test cases. This has been suggested by, for example, Phadke (1997). He calls this type of software testing 'Robust Testing'. The usage space (or input space supplemented with the situational noise factors) is parameterized in such a way that testing could be performed according to a two- or three-level fractional factorial design. A similar approach has been used by Ericsson Microwave in testing software for radar applications; see Berling (2003) and Berling and Runesson (2003).

13.2.2 Maintained Reliability

For long term reliability, maintenance issues are important. Even if a great deal is done in the product development phases to reduce the occurrence of failures, inevitably, failures do develop during usage. Now the problem is how to catch a growing potential failure before there are any consequences. This is the subject matter of Reliability Centred Maintenance (RCM) and Condition Monitoring (CM). The RCM concept was introduced in the Aerospace industry in the 1970s by Nowlan and Heap (1978); see also Bergman (1985). In the early days of maintenance planning in the aerospace industry, all equipment was subject to preventive maintenance – old equipment was replaced regularly by newly maintained equipment. Nowlan and Heap (1978) found, on the basis of extensive data obtained from the operations of United Airline, that much equipment had a failure rate that was either decreasing or constant, i.e. modelled as the first and/or second phases of the bath-tub curve failure rate model (see Figure 1.1). However, for such equipment, preventive maintenance is only costly and no reliability improvements are achieved. In this spirit, Nowlan and Heap (1978) suggested a systematic scheme for the selection of maintenance activities avoiding costly and unnecessary preventive maintenance replacements. An important part of that scheme was the utilization of CM. In CM some indicators correlating with the deterioration of the equipment are sought. By monitoring these indicators it should be possible to find failure modes of deteriorating equipment before they have developed to a stage where there is a considerable risk of failure. Naturally, this gives rise to an optimization problem which has been studied extensively, see for example Bergman (2007) and references therein.

Reliability Centred Maintenance has spread far beyond the aerospace industry and is now applied in a great variety of industries, for example the nuclear, offshore, manufacturing and hydropower industries; for comprehensive descriptions of RCM see, for example, Andersson and Neri (1990) and Moubray (1997).

13.3 Organizational Perspectives on Reliability

Reliability and safety is not only about singular components and parts breaking and thereby leading to catastrophic failures – it is more often about multiple events occurring in complex systems operated in complex organizational settings. We will indicate some perspectives taking this into account. The first one is rather pessimistic, but provides necessary insights before more optimistic perspectives can be advised.

13.3.1 Normal Accidents

In a seminal book by the sociologist Charles Perrow, a number of disastrous events were studied, including the Three Mile Island Nuclear Power plant accident, petrochemical plant events, aircraft and airway events, marine events, etc. Perrow (1984) draws the conclusion that accidents are prone to appear in complex systems – they could in a way be seen as 'normal'. He found that the catastrophic end results were the combined system effects of apparently minor interdependent events in a complex tightly coupled technological settings. In many complex systems the following points might be realized:

1. Failure in one part (due to design, equipment, procedures, operators, suppliers and material, or environment – all six abbreviated as the DEPOSE components) may coincide with failures in a completely different part of the system. Such unforeseen combination of failures may cause a further cascade of failures.
2. Such failure cascades may give surprising effects – the logic of system failure is very different from functional logic. Redundancies may be reversed and eliminated and safety devices might be sources of disastrous events.
3. In essence, there are unlimited possibilities for such failure combinations.
4. Due to accelerating failure processes operators may lose control and take actions making things even worse.

Perrow's conclusion is that accidents are inevitable in complex and tightly coupled systems – they are 'normal'!

The ideas put forward in the earlier chapters of this book, especially in Part One, and applied to a system level is one means to reduce complexity and loosen tight couplings. We shall briefly mention two other perspectives in the same direction, but much further into the organizational realm: 'Resilience Engineering' and 'High Reliability Organizations'.

13.3.2 Resilience Engineering

In resilience engineering the focus is on the design of a system such that failures, which inevitably occur but in unexpected ways, are taken care of in a safe way and, by design, no disastrous events are allowed to occur. Failures, however unexpected, lead to safe states rather than an accelerating cascade of new failures. In a way this may be seen as a contradiction – at least when safety is regarded as a property that a system *has* rather than something the system *does*, see Hollnagel and Woods (2006). These authors emphasize that a resilient system should have such qualities that problems are anticipated before they arrive, the system is constantly watchful, i.e. attention is given to deviations which may develop to failures, and they should have a preparedness to respond to signals indicating potential problems. This means that 'a resilient system must have the ability to anticipate, perceive, respond' (Hollnagel and Woods, 2006). In Hollnagel et al. (2006) the concepts and precepts of resilience engineering are discussed. It is a relatively new area with a number of similarities to our next topic.

13.3.3 High Reliability Organizations

An interesting and a very different perspective from the ones studied above is suggested by Karl Weick and his co-workers. Consider an aircraft carrier with a large amount of take-offs and landings corresponding to a large International Airport but with a size of only a few football fields, and crowded with dangerous ammunition, flammable fuel, nuclear power generators, etc. A highly risky environment and, still, very few accidents occur. This was the starting point for an interesting discourse on High Reliability Organizations (HRO) – organizational culture rather than technical issues was found to be at the core of these organizations, see e.g. Roberts and Bea (2001) and Weick and Sutcliffe (2007), and references cited therein. The latter book 'Managing the Unexpected – Resilient Performance in an Age of Uncertainty' provides an interesting discussion on the principles of HRO. We will provide only a very short introduction here. A central concept is that of mindfulness – meaning that people in the organization are highly aware of the possibility that small things may, in the future, develop into something dangerous. Discriminating between details, they single out those deviating from the expected, indicating something that may go wrong. On the basis of qualitative research, the authors and their companions have extracted some principles in the use in HROs. The following principles are discussed by Weick and Sutcliffe (2007).

1. *Preoccupation with failure*: any lapse is taken as a symptom of a failure, and it is realized that small errors might pile up and result in something dangerous in combination; small incidents are reported and failure indications are never ignored; learning from mistakes is considered very important and, continuously, possible mistakes are assessed and remedies are taken.
2. *Reluctance to simplify*: failures often occur due to the high complexity of the system – this was strongly emphasized by Perrow (1984) as discussed above – and the increasing complexity will make severe accidents commonplace. Simplification is natural, and in simple systems cause–consequence chains are easy to follow. That is *not* the case for complex systems. The full picture should be taken into account.
3. *Sensitivity to operations*: it is in the operations, the very situation where things happen, where the first signs of anomalies do occur; the Japanese tell us to 'go to the gemba', i.e. be sure that you have a good understanding of the operations.
4. *Commitment to resilience*: failures do occur in all systems, also in HROs. However, they are organized to be able to cope with these failures to limit their consequences and quickly come back to a safe state. Note that this point is in itself the origin of a perspective on reliability as discussed above.
5. *Deference to expertise*: in HROs, decisions are taken where most of the knowledge is to be found – out in the peripheries of the organizational structure. Rigid hierarchies where decisions are taken at the top of the organization are too slow to cope with quickly upcoming signs of growing disasters. However, as Weick and Sutcliffe emphasize, experience is not always a warranty for expertise – learning has to be emphasized.

The above discussion is just a short reminder that in complex systems not only technical aspects have to be taken into account, but also should cultural ones. We refer the reader to the book by Weick and Sutcliffe (2007) and references given therein.

13.4 Industrialization of Robust Design Methodology

After this detour into some alternative perspectives on reliability we will come back to our main theme, but, from a different angle. The ideas emphasized in this book have to be adhered to in the design of the systems (products or processes). Thus the product development process has to be changed in a way such that these ideas are realized into the developed products or systems. This also requires some kind of cultural change. We will give some brief comments based on a Japanese case and some cases from our own experience.

13.4.1 Adoption of Robust Design Engineering at Fuji Xerox

For a long time, with the first applications as early as 1972, Fuji Xerox has addressed the question of robustness utilizing the ideas suggested by Dr Genichi Taguchi as early as the

1950s. They use the name Quality Engineering (QE) for, essentially, the same process we here have called Robust Design Methodology. They make explicit reference to reliability; from their applications it is obvious that reliability improvement has been a very important outcome of their endeavours. However, in spite of their long experience, the application of these ideas as a normal procedure in product development has not been a self-evident business. Thus, they have made thorough investigations concerning key success factors. The following factors have been found to be decisive for the success of their approach, see Saitoh et al. (2003a, 2003b; note that their QE is equivalent to RDM in our terminology):

1. Policy announcement and involvement by the top of the department.
2. Identification of person in charge of promotion and establishment of promotion organization for QE.
3. Promotion activities in accordance with the QE application status of the department.
4. Continuous training and establishment of a mechanism for making engineers use the QE.
5. Clarification of themes to which QE will be applied and management of the progress.
6. Continuously held meetings for discussion and application.
7. Use of internal or external consultant/expert.
8. Clarification of result of application. (Barometer of success and actual result.)

At Fuji Xerox there is an impressive number of applications of QE. As an example, there were 390 applications in the year 2005 (from Tatebayahshi, September 2006). A special Promotion Committee has been given the responsibility to promote the use of QE; they particularly emphasize the improvement of the key success factors described above. The emphasis is not only on reliability improvements but also on reducing time-to-market.

13.4.2 A Swedish Case

At one company, the name of which we choose not to disclose here, one of our PhD students has been working on the introduction of RDM. Also a number of master's theses have been utilized in the development process to produce insights into the difficulties and possibilities to introduce RDM. In the company great efforts have been made to get a RDM in place to decrease warranty claims and manufacturing problems. Their first experiences made clear the importance of having a systematic process for elicitation of customer needs and wants in place. It is very important that the RDM efforts are made on product characteristics which are of importance to the customers, so-called Key Product Characteristics (KPCs). It is only for this type of characteristic that it is important to make efforts to create insensitivity to noise. Another important learning outcome was the importance of the adaptation of the methods to their own engineering environment – a consultant-led initiative was not as successful as it was hoped to be due to the limited adaptation to the engineering culture of the company. The experience also supports the kind of findings made by Fuij Xerox as discussed above.

13.4.3 The SKF Case

In 2005, SKF (see footnote 1, Chapter 1) started a Six Sigma programme initiated by their Chief Executive, Tom Johnstone. The results of this programme, emphasizing variation and its reduction for important product and process characteristics, have been very good. Their Annual Report for 2007 states:

> One key focus activity for SKF over the past three years has been the introduction of Six Sigma. This is going very well and today we have 378 Black Belts. Last year we recorded over SEK 300 million in hard savings for our Six Sigma projects, up by 50% from 2006. We are also using Design for Six Sigma more and more in the Group and widening this work to also include customer programmes and customer training.

For a company with a good result from Six Sigma activities, i.e. the systematic application of statistical problem solving methodology to reduce variation and thereby increase quality, it is natural to continue into the earlier stages of the development of products and processes; i.e. start to work with Design for Six Sigma. In short, Design for Six Sigma (DFSS) means that product and process characteristics of importance to customers and manufacturing (so-called Key Product/Process Characteristics, KPC) are selected, noise factors influencing these characteristics are found and the product or process is made insensitive to the variation of these noise factors. Both creative means and systematic parameter design methodologies are utilized, see Chapter 1. In many companies the DFSS process is applied to selected products and processes utilizing a special procedure similar to the stepwise procedure of Six Sigma (where the steps are called define, measure, analyse, improve and control; for a more complete description see Magnusson et al., 2003). In DFSS the corresponding steps have different descriptions, with one of the more common ones being:

1. *I – Initiate project*. Initiate and prepare the project, establish the business case and project the infrastructure.
2. *D – Define requirements*. Gather and understand customer requirements.
3. *D – Develop concepts*. Develop, evaluate, synthesize and select design concepts.
4. *O– Optimize design*. Model, analyse and develop design to optimize product and process capability.
5. *V – Verify and validate*. Test design outputs against customer requirements under defended operating conditions.

At SKF, it was decided to integrate this process into the existing product development process as described in Figure 13.1. At each stage of the product development process, non-binding recommendations are given concerning the tools which could be utilized to meet the deliverables needed to go further to the next stage. The tools are just recommendations and should not limit the creativity and imagination of the designer. However, checklists are given in order to capture all the outputs needed at each stage. From a DFSS perspective, the checklists ensure that nothing important is overlooked or neglected. We shall not go further into details – for the interested reader we recommend Hasenkamp and Ölme (2008) and Hasenkamp (2009), where more detailed descriptions are found.

13.4.4 Standardization

It has been noted that in the business-to-business market there is a need for buyers to put requirements on their system suppliers to apply RDM in the product development processes

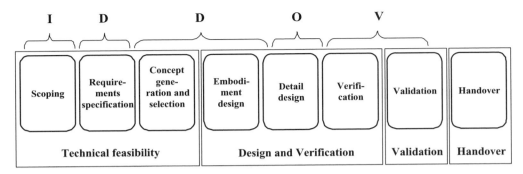

Figure 13.1 The product development process at SKF and its relation to steps in DFSS (adapted from Hasenkamp (2009)).

in order to have some assurance that reliability improvement measures have been taken. This requires some kind of standardization concerning concepts and procedures. A workshop on the necessity of international standardization to accelerate new technology development with statistical and related methods was held in October 2008 at the the ISO TC 69 meeting (ISO is the International Standards Organization and TC69 is its Technical Committee handling quality and statistical methods). The objective of the workshop was to clarify mission and scope of a possible new subcommittee for standardizing statistical methods to accelerate new product and technology development including Quality Function Deployment (QFD), robust parameter design, implementation of customer satisfaction survey, etc., in TC69.

13.5 Adoptions of Fatigue Reliability Methodology

In line with the proposed approach, unwanted variation is considered as the cause of unreliability. In order to implement this way of thinking the sources of variation have to be explored and quantified. Systematic errors also have to be modelled as random variables to permit a unified approach to all kind of disturbances, deviations, errors and failures.

As T.P. Davis at Ford pointed out the classic probabilistic approach to use the tail in the distribution of life requires much more data than are usually available. Accordingly, our proposed approach is based on second moments reliability assessments, which are more attainable than tail estimations.

13.5.1 Adoption at Volvo Aero

To use the approach in industry, however, a design code has to be formulated. In collaboration with us, Volvo Aero produced such a code and applied it in its design work. In one application it was observed that the fatigue model had to be improved to obtain predictions with sufficient performance. In another application a narrower tolerance was required. The strength of the method is that it clearly indicates where the improvements will bring the greatest gains. Sometimes the information available is too poor to allow inference with sufficient precision, sometimes a modification of the specifications is needed. The adoption of the methodology at Volvo Aero is promising.

In the Volvo Group there is much experience in using probabilistic design methodologies and therefore there is a good base for adopting the current approach and the ideas are going to be tested. Other branches of the mechanical industry have also shown an interest in this.

13.5.2 Stone Cladding Reliability

A seemingly totally different topic is the problem of dimensioning stone cladding. A recent EU project says: 'As stated above, many uncertainties in the available assessment models for stone cladding remain unresolved. Standard test procedures are not sufficient to get enough knowledge about the strength, effects of size, environment, fatigue and surface conditions are only partly investigated. All these uncertainties must be taken into account in a reliability assessment for safe constructions. At present this is achieved using the Eurocode method of partial safety factors. The conclusion of the findings in this project, however, gives rise to the demand for a better reliability approach. Initial studies in the I-STONE project suggest that a *second moment reliability method* should be a good trade off between actual knowledge and safety demands, and the development of such an approach would be a natural continuation of this project'.

13.5.3 Load Analysis for Trucks

In the process of exploring the different sources of variation for fatigue life assessment it is crucial to study the load variations. Together with us the European truck industry runs a joint project to produce a guide for load analysis where exploration of the sources of variations forms the basis for future measurement campaigns. There are different kinds of trucks, different missions, different classes of roads, different topographies, different national regulations, different hauliers and different drivers belonging to the same haulier.

13.5.4 Strength Variation

Another source of variation is the material strength. For hard clean steel the fatigue limit is governed by the size of the largest inclusion in the loaded part of the construction. Now inclusions are rare, which makes the variation large. It is important for both the manufactures of steel components and for the steel producers to be able to estimate the strength variation. One way to proceed is to estimate the size and the frequency of the nonmetallic inclusions. Recently there was an ASTM standard proposed based on measuring inclusion sizes on the surface of polished specimens. Another approach is to use ultrasonic tests or fatigue tests.

13.5.5 Summary

In our approach the idea is to estimate, from the very beginning, the sources of variation and their sizes based on previous experience. The first stage estimates are used to direct the efforts for improvements to the sources that are most important.

13.6 Learning for the Future

The cooperation leading to this book has been most rewarding. Three different research groups have joined in a reliability research project supported by the Swedish Strategic Research Foundation via the Gothenburg Mathematical Modelling Centre. The three organizations, all within an affiliation to Chalmers University of Technology, are the Division of Mathematical Statistics (Department of Mathematical Sciences), Quality Sciences (Department of Technology Management and Economics) and the Fatigue Group at Fraunhofer–Chalmers Centre for Applied Mathematics (FCC). The cross disciplinary work resulting from this cooperation has given new dimensions to the research performed by the different groups, and new relationships have been created. We are looking forward to future research cooperations based on the platform now created.

Some interesting developments of the ideas given in this book on RDM for reliability improvement are the following:

- advancement of on-board - diagnostics
- robustness in a maintenance perspective
- advancement of robust testing in software development
- design and analysis of experiments for identification of robust design solutions
- fatigue and robustness considerations in nanotechnology.

In the course of writing of this book we have also realized that there is a need for a textbook on the themes advanced here. That book should include time dynamic models and merge the ideas behind the Six Sigma approach and the VMEA methodology.

References

Andersson, R. T. and Neri, T. *Reliability Centered Maintenance, Management and Engineering Methods*. Elsevier, Barking, 1990.

Bergman, B. On reliability theory and its applications (with discussion). *Scandinavian Journal of Statistics*, **12**: 1–42, 1985.

Bergman, B. Stationary replacement strategies. In *Encyclopedia of Statistics in Quality and Reliability*. John Wiley & Sons, Ltd, Chichester, 2007.

Berling, T. *Increasing product quality by verification and validation improvements in an industrial setting*. PhD thesis, Department of Communication Systems, Lund University, Lund, 2003.

Berling, T. and Runesson, P. Efficient evaluation of multifactor dependent system performance using fractional factorial design. *IEEE Transactions on Software Engineering*, **29**: 769–781, 2003.

Hasenkamp, T. *Designing for robustness*. PhD thesis, Division of Quality Sciences, Chalmers University of Technology, Gothenburg, 2009.

Hasenkamp, T. and Ölme, A. Introducing design for Six Sigma at SKF. *International Journal of Six Sigma and Competitive Advantage*, **4** (2): 172–189, 2008.

Hollnagel, E. and Woods, D. D. Epilogue: Resilience engineering precepts. In *Resilience Engineering, Concepts and Precepts*, E. Hollnagel, D. D. Woods, and N. Leveson (eds), pp. 347–358. Ashgate, Burlington, 2006.

Hollnagel, E., Woods, D. D. and Leveson, N. *Resilience Engineering, Concepts and Precepts*. Ashgate, Burlington, 2006.

Magnusson, K., Kroslid, D. and Bergman, B. *Six Sigma – The Pragmatic Approach*. Studentlitteratur, Lund, 2003.

Moubray, J. *Reliability Centred Maintenance*, 2nd edn. Butterworth–Heinemann, London, 1997.

Nowlan, F. S. and Heap, H. F. *Reliability Centred Maintenance*. Report AD/A066-579, National Technical Information Service, US Department of Commerce, Springfield, VA, 1978.

O'Connor, P. D. T. *Practical Reliability Engineering*, 4th edn. John Wiley & Sons, Inc., New York, 2002.

Perrow, C. *Normal Accidents, Living With High-Risk Technologies*. Basic Books, New York, 1984.

Phadke, M. S. Planning efficient software tests, crosstalk. *Journal of Defense Software Engineering*, **10**(10): 11–15, 1997.

Roberts, K. H. and Bea, R. Must accidents happen? Lessons from high reliability organisations. *Academy of Management Executive*, **15**: 70–81, 2001.

Runesson, P. and Wohlin, C. Cleanroom software engineering in telecommunication applications. *SEAT'93 – Proceedings Software Engineering and its Applications in Telecommunications*, Paris, December, pp. 369–378, 1993.

Saitoh, K., Yoshizawa, M., Tatebayashi, K. and Doi, M. A study about how to implement quality engineering in research and development (Part 1). *Journal of Quality Engineering Society*, **11**: 100–107, 2003a.

Saitoh, K., Yoshizawa, M., Tatebayashi, K. and Doi, M. A study about how to implement quality engineering in research and development (Part 2). *Journal of Quality Engineering Society*, **11**: 64–69, 2003b.

Tatebayahshi, K.. Introduction and implementation of quality engineering (QE), in Fuji Xerox. *Presentation at a Robust Design Methodology Conference*, Chalmers University of Technology, Gothenburg, September 2006.

Weick, K. and Sutcliffe, K. M. *Managing the Unexpected, Resilient Performance in an Age of Uncertainty*. Jossey Bass, San Francisco, 2007.

List of Abbreviations

AGREE	Advisory Group on Reliability of Electronic Equipment
APQP	Advanced Product Quality Planning
CM	Condition Monitoring
DEPOSE	Design, Equipment, Procedures, Operators, Suppliers, Environment
D-FMEA	Design Failure Mode and Effects Analysis
DFSS	Design for Six Sigma
DOE	Design of Experiment
ENIAC	Electronic Numerical Integrator and calculator
ETA	Event Tree Analysis
FEM	Finite Element Method
FMEA	Failure Mode and Effects Analysis
FOSM	First Order Second Moment
FTA	Fault Tree Analysis
GLM	Generalized Linear Model
HAZOP	Hazard and Operability Study
HRA	Human Reliability Methods
HRO	High Reliability Organizations
KPC	Key Product Characteristics
MCTP	Markov Chain of Turning Points
MLC	Main Life Cycle
MOM	Method of Moments
P-FMEA	Process Failure Mode and Effects Analysis
QE	Quality Engineering (often used in Japan to denote Robust Design Methodology)
QFD	Quality function Deployment
RBD	Reliability Block Diagram
RCM	Reliability Centred Maintenance
RDM	Robust Design Methodology
RPD	Robust Parameter Design
RSM	Response Surface Methodology

Robust Design Methodology for Reliability: Exploring the Effects of Variation and Uncertainty
edited by B. Bergman, J. de Maré, S. Lorén, T. Svensson
© 2009, John Wiley & Sons, Ltd

S-FMEA   System Failure Mode and Effects Analysis
SPC   Statistical Process Control
VMEA   Variation Mode and Effects Analysis
VRM   Variation Risk Management
VRPN   Variation Risk Priority Number

Index

Robust Design Methodology for Reliability: Exploring the Effects of Variation and Uncertainty
edited by B. Bergman, J. de Mare, S. Lorén, T. Svensson
© 2009, John Wiley & Sons, Ltd